SNOWY

BRAD COLLIS

SNOWY

THE MAKING OF MODERN AUSTRALIA

HODDER & STOUGHTON
SYDNEY   AUCKLAND   LONDON   TORONTO

ACKNOWLEDGEMENTS

The author gratefully recognises the assistance of the Snowy Mountains Hydro-Electric Authority, in particular the following: Mr Robin Binder, Ms Jill Blackburn, Mr Neville Phee, Mr Frank Rodwell, Mr Russell Stores, Mrs Elizabeth Langford, Mr Errol Cutler, Mr Clive Daly and Mr Steve (Smokey) Perkins; also Mr Doug Price, Lady Hudson, Mr Tom Lewis, Mr Mike Bender, Mr Clem Mackay, Mr Keith Goodwin, Mr Lindsay Hain, Mr Julian Cribb for his literary guidance; the staff of the Australian Archives in Canberra and Melbourne, whose cheerful assistance greatly eased the research task. The National Library Canberra and *The Age* Melbourne have kindly supplied photographs as indicated. All photos not otherwise acknowledged are included by courtesy of The Snowy Mountains Hydro-Electric Authority or generous private collectors.

First published in 1990
by Hodder & Stoughton (Australia) Pty Limited,
10-16 South Street, Rydalmere, NSW, 2116.

© Brad Collis, 1990

This book is copyright. Apart from any fair dealing for
the purposes of private study, research, criticism or
review as permitted under the Copyright Act, no part may
be reproduced by any process without written permission.
Enquiries should be addressed to the publisher.
National Library of Australia Cataloguing-in-Publication entry

Collis, Brad, 1955- .
  Snowy.

  ISBN 0 340 49640 1.
  ISBN 0 340 53385 4. (limp edition)

  1. Snowy Mountains Hydro-Electric Scheme—History. 2.
  Hydroelectric power plants—New South Wales—Snowy
  Mountains—History. 3. Snowy Mountains Region (N.S.W.)
  —History. I. Title.

621.31'2134'09944
Typeset by G.T. Setters Pty Limited, Kenthurst, NSW
Printed in Hong Kong by South China Printing Company (1988) Limited
Designed by Stan Lamond, Lamond Art & Design, Coogee, NSW

Previous pages

Excavation for the Tumut-1 power station, October 1957.

Senator William Spooner (left) with an Australian-born foreman Bill Simpson (second from right) and German-born tradesmen at Adaminaby, 1950.

Opposite

Drillers aboard a three-deck drilling jumbo in the Tumut-2 tailrace tunnel.

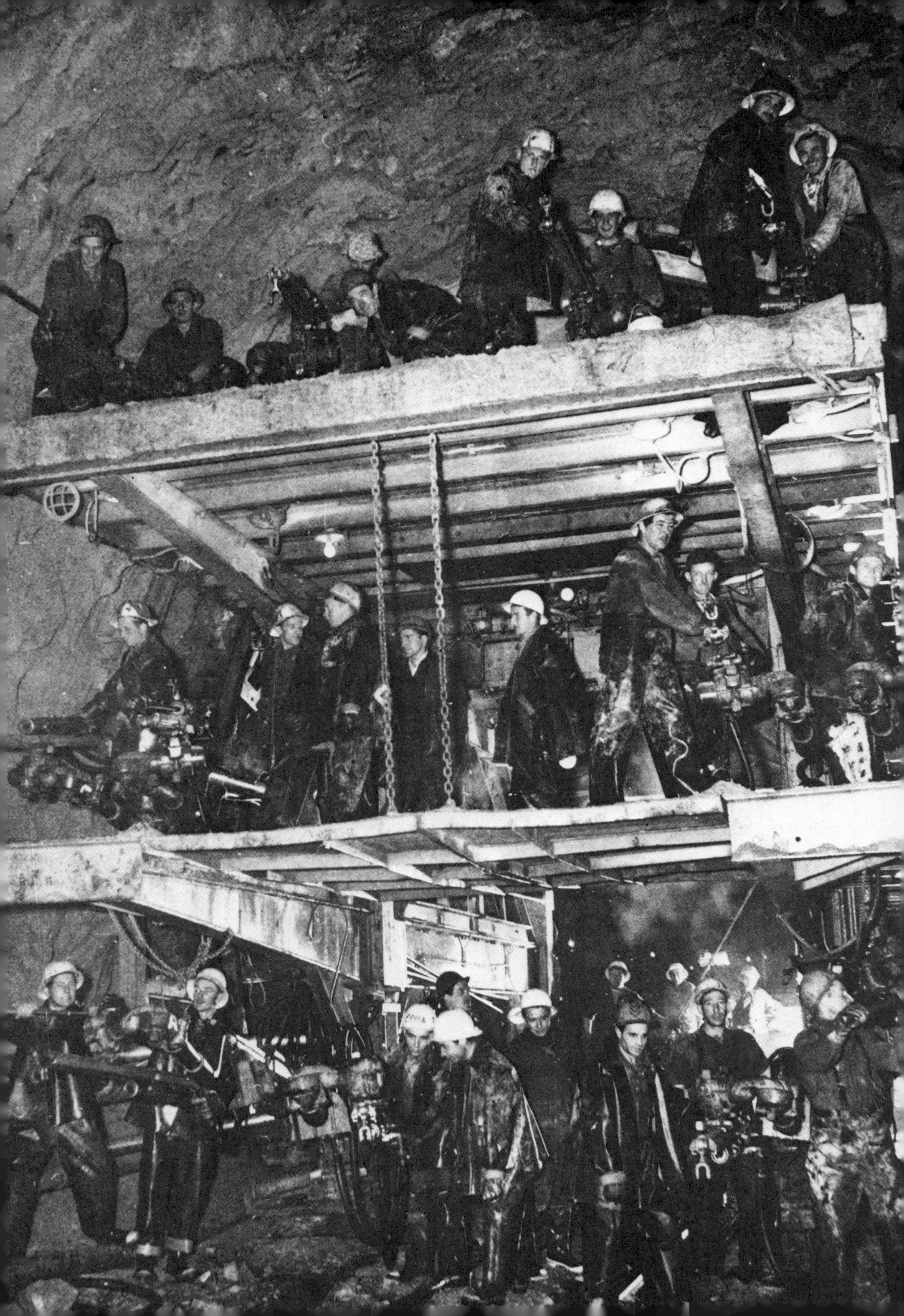

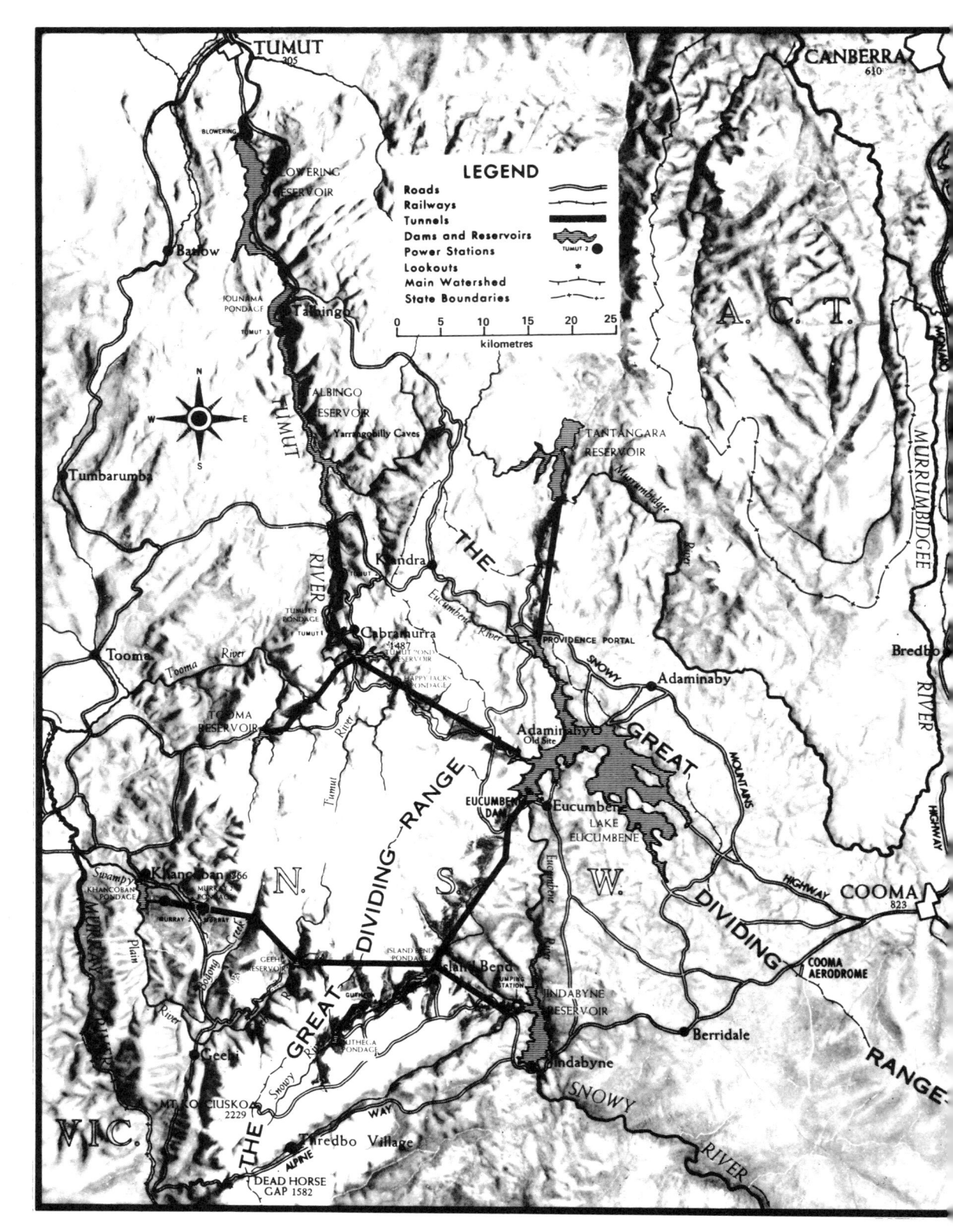

THE SNOWY MOUNTAINS AREA

CONTENTS

FOREWORD 8

Prologue
AMONG FREE PEOPLE: KOSCIUSZKO AND PAUL STRZELECKI 10

Chapter One
THE MAJOR 16

Chapter Two
POLITICS OF WATER 30

Chapter Three
ON A WING AND A PRAYER 46

Chapter Four
JINGLE OF BIT, CREAK OF LEATHER 68

Chapter Five
CONVERGENCE 86

Chapter Six
TENTATIVE BEGINNINGS 118

Chapter Seven
BUSH COOK 148

Chapter Eight
TUNNELS OF BLOOD 160

Chapter Nine
NEW AUSTRALIANS 188

Chapter Ten
THE LAWMEN 212

Chapter Eleven
WINTER 230

Chapter Twelve
FLOODS AND REBELLION 246

Chapter Thirteen
AFTERMATH 264

Epilogue
FUTURE DIRECTIONS 282

Appendix One
ROLL OF HONOUR 286

Appendix Two
TECHNICAL INFORMATION 289

INDEX OF PEOPLE 300

GLOSSARY OF ABORIGINAL NAMES 304

FOREWORD

The Snowy Mountains Scheme changed Australia forever. A country founded on stolid British stock almost overnight became one of the world's great pancultures. Tens of thousands of workers from more than thirty different lands poured into what was once the undisturbed pastoral realm of the high country stockmen, the southern Outback.

As they drilled and tunnelled into the mountains their energy and skills gave the country a mighty push into the twentieth century. The vibrant interaction of their many diverse cultures and the scale of the project in terms of technology made the Scheme one of the seven engineering wonders of the modern world.

The mountains still guard the bones of scores of men who lost their lives in the effort. But forty years on those who remain to look back do so with pride in their share in the Scheme and its contribution to the development of Australia.

Most came from war-torn Europe and many had been recent and bitter enemies. But by and large they left their hatreds behind, in their determination to grasp the blessed opportunity of a second chance. Dispossessed by war, they worked with great energy and entrepreneurial drive to better both themselves and their new homeland.

The building of the Snowy Mountains Scheme put Australia on the forefront of world construction technology. An incredible feat for a young country with an economy still based almost entirely on agriculture. If ever there was a national monument from which a people could draw pride and confidence it is the Snowy Scheme. It has been described in other countries as a monument to political vision, social harmony and engineering excellence—yet a generation of Australians has grown up barely conscious of its existence.

Perhaps this is because it has operated faultlessly and without fanfare for more than two decades and perhaps also because its most extraordinary features are hidden beneath the mountains—out of sight of the hundreds of thousands of people who annually flock along its access roads to the ski slopes.

The Snowy Scheme was the product of people with vision, from a time when people had the courage to dream and build on those dreams. The imagination and strength of Prime Minister Ben Chifley, his Minister for Works, Nelson Lemmon, and Sir William Hudson, the engineer, gave wings to a vision which emerged in the nineteenth century and which is now projecting into the twenty-first century. As humanity faces an uncertain future from pollution, unsustainable agricultural practices and global warming under the so-called Greenhouse effect, hydro-electricity remains one of the cleanest renewable sources of power.

During the two decades of construction, the numerous vast projects making up the Scheme overlapped and confused the overall picture. It was like a gigantic jigsaw; lots of pieces, but no picture. Because only they understood both the pieces and the picture, the brilliance of the engineers who designed and co-ordinated it all was overlooked.

It must be emphasised, however, that the Scheme was and is one of the world's greatest engineering achievements. Its two main developments, one based on the rivers feeding into the Murrumbidgee River, the other on those feeding the Murray River, involved the blasting of 12 tunnels through the mountains, the building of 16 dams to hold the water and feed it through 7 power stations, 2 of them deep under the mountains.

To the thousands who came to build the Snowy and the tens of thousands who have migrated since, the Snowy is the place where they were given a chance to start again, a place to build a new country based on shared prosperity and social justice, a country open to new ideas and lateral thinking; a country which cares for its environment and the welfare of future generations; a country which because of its diverse ethnic mix, is tolerant and compassionate.

The Snowy Mountains Scheme was a testing ground for Australia's technical capacity, will to achieve and ability to absorb social change. It is important for us all that the men who died on and inside the Snowy Mountains did not fall in a useless war but in the building of a new and vibrant nation—their nation. The migrants who were drawn to Australia by a search for new hope and peace, discovered it in the Snowy Mountains...

In a quiet street not far from where he worked at the Polo Flat workshop in Cooma, a former Luftwaffe pilot enjoys his retirement in endeavouring to perfect the arts of calligraphy and home wine-making. One of his closest friends during his working years on the Scheme was a Polish bomber pilot operating over Germany from exile in England. In another street a former Australian infantryman had as his neighbours a former U-Boat commander and a vanquished Polish soldier. At one stage during the war the Pole had been a prison camp guard in Scotland where the captured U-Boat captain had been interned. Now they are neighbours and friends in a new country on the other side of the world.

Numerous similar stories are part of the Snowy saga and such people worked hard to become Australians... In their turn native-born Australians have seen their cultural and social life immeasurably enriched by the contribution of the strangers from strange lands who first flooded in to work on the Snowy Mountains Scheme.

Over the years a number of books have been written about the Snowy but due probably to the narrowness of time since the Scheme's construction, they focused on specific aspects such as its technical features or a particular migrant group.

My aim has been to draw all the threads together; to offer a picture of the Scheme as a whole... its people, its political and industrial dramas, the tragedies, engineering achievements, and most importantly its role in the shaping of modern Australia.

—Brad Collis

PROLOGUE

The 6.5 km Tumut–2 tailwater tunnel which returns water used by the Tumut–2 power station to the Tumut River.

AMONG FREE PEOPLE: KOSCIUSZKO AND PAUL STRZELECKI

In the summer of 1839–40, drought was very much on the mind of a prominent sheep grazier, James Macarthur, son of Australian wool industry pioneer, Captain John Macarthur. He had put 500 pounds into an expedition to find new country for the merino flock the family was continuing to build up at Camden, near Sydney. The flock would one day be the foundation of one of Australia's biggest industries, wool, of which Australia would also become the world's foremost producer—but that was still far into the future.

For the moment, on 11 March 1840, James Macarthur was simply a worried and mildly annoyed man. Worried by the debilitating effects of a land that was always parched; and frustrated at finding himself edging painfully up the side of a mountain following two native guides and a bobbing backpack strapped to the broad shoulders of a wandering Pole, Paul Strzelecki. Macarthur was impatient to find new pastures, not conquer mountains. As he followed the Pole's measured steps through thick scrub and up steep faces he wondered, not for the first time, how he had allowed himself to be talked into leaving the main party in the upper Murray Valley for this unplanned excursion to the topmost peaks of the Australian Alps. His only interest was his flocks, which were multiplying too fast for the available feed.

He had been lured, as he wrote later, 'by the ardour of discovery to ascend the highest known peak in Australia' and was now regretting it. The terrain at times was so steep that the men could climb only by grasping shrubs and small saplings. The weather was very hot, time was short and they were forced to push on under moonlight.

After five hours climbing from dawn on the second day they came to a natural clearing, a camping ground used by Aborigines during their annual feast on the bogong moth. They were about 1300 metres (4000 feet) above the Geehi River. The valley below was filled with a dense, white mist. After several more hours of struggling up the slope they crossed the snowline. A cold fog settled on the men and they sent the two Aborigines back for blankets and food, with instructions to set up a camp below the snow.

The two Europeans continued on alone until they reached a plateau covered with alpine plants and a boggy marsh—the source of the many streams that fed the rivers which emerged from the alpine plains. The bog was (and still is) a sphagnum moss community containing primitive alpine plant species from the last Ice Age. The men then struck south for several kilometres until Strzelecki stopped, lowered his pack of instruments to the ground and pointed. Even Macarthur later admitted it was a grand sight. The view encompassed about 12 000 square kilometres of south-eastern Australia.

They had arrived on the summit of what was later named Mount Townsend. Around them were several peaks and, using his instruments, Strzelecki determined that the highest of them was a rounded cone of bare granite about five kilometres away on the other side of a ravine. He pointed to the cone-shaped peak: 'That is the highest... most likely the highest point of the continent. It reminds me of the tumulus over the tomb of my countryman, Tadeusz Kosciuszko, in Krakow. He fought and died for Poland's freedom; therefore I name this Kosciuszko because here in this foreign country I am among free people.'

They were prophetic words. Just over a century later thousands of his compatriots would come to know these mountains and express the same heartfelt sentiment. It was perhaps also fitting that his companion was Macarthur who, as an emerging politician,

James Macarthur.
PHOTO: NATIONAL LIBRARY CANBERRA

Paul Strzelecki.
PHOTO: NATIONAL LIBRARY CANBERRA

Thaddaeus Kosciuszko.
PHOTO: NATIONAL LIBRARY CANBERRA

would later play a leading part in ending the colony's role as a dumping ground for England's crowded gaols.

Strzelecki, an enthusiastic and persuasive man in his early thirties, had left his Polish hometown Poznan to rove and explore, reputedly in an effort to overcome the frustration of a thwarted attempt to elope with the girl he was madly in love with. Her parents had considered him unsuitable to marry their cherished daughter. He came from a poor gentry family without land or title, though somewhere between leaving Poland and arriving in Australia he assumed the title of Count, which smoothed his entry into the colony's circle of influence. Strzelecki had a keen interest in mineralogy and geology and had arrived in Sydney after a circuitous journey through England, North America, South America and New Zealand. Geology was an infant science at the time and Strzelecki, like many others in the field, was self-taught. Nevertheless, he was credited with discovering copper in Canada during his journey through that country.

His sweetheart in Poland, Aleksandryna Turno, featured much in his writings and he regularly recounted his exploits to her. On his return from the expedition with Macarthur, he wrote to her ecstatically:

> Here is a flower from Mount Kosciuszko—the highest peak of the continent, the first in the new world bearing a Polish name. I believe that you will be the first Polish woman to have a flower from that mountain. Let it remind you ever of freedom, patriotism, and love...the highest peak of the Australian Alps. It towers over the entire continent, which before my coming, had not been surmounted by anyone. With its everlasting snows, the silence and dignity with which it is surrounded. I have reserved and consecrated it as a reminder for future generations upon this continent of a name dear and hallowed to every Pole, to every human, to every friend of freedom and honour—Kosciuszko.

The love affair, by correspondence, between Paul Strzelecki and Aleksandryna Turno—or Adyna, as he called her—lasted for twenty-two years. The affair remained secret, each letter avowing undying love, yet they remained barred from marriage by the social division imposed by her parents. Neither ever married: Strzelecki sought reward and comfort in adventure and science; and she shared his experiences through his many letters and writings.

Macarthur also recorded the ascent in his diary: 'I could not but respect and feel deep sympathy with my friend when, with his hat off, he named the patriot of his country.' He decided to leave Strzelecki to himself in his moment of triumph and made his way back down the mountain to where he had ordered the Aborigines to make camp. Strzelecki chose to continue on to the peak he had just named, to stand atop the highest point on the land, alone with his thoughts.

From their 'diversion' to become the first Europeans to stand on the roof of Australia, Strzelecki and Macarthur later made their way back down to the Murray Valley, through the Nariel Valley and on to an exploration of Gipps' Land (the Gippsland region of Victoria). There, looking through a newcomer's eyes at the debilitating effect of drought in south-eastern Australia, Strzelecki penned a perceptive warning of the link between drought and colonisation. It seemed to him that the drought might be worsened by the fact that plants, forest and undergrowth which normally sheltered the soil and conserved moisture had disappeared 'under the innumerable flocks and axes which the settlers had introduced'. He looked ahead to the need for 'the rotation of crops, dams to arrest the torrents, reservoirs to contain and preserve them, artesian wells...and irrigation'. In a report to the British Parliament, Strzelecki wrote: 'Irrigation becomes the first measure with which the agricultural improvements of Australia must begin.'

Paul Strzelecki, the wandering Pole, later became a British subject and was knighted by Queen Victoria for his services to exploration in Australia. He died peacefully in his sleep in London in October 1873. The trek he had taken with Macarthur was a minor human chapter in the endless cycle of seasons on a dry continent, but it strengthened the thread that was already beginning to bind the future of Australians to these mountains.

CHAPTER ONE

Major Clews near Scammell's lookout in 1954.

THE MAJOR

THE MAJOR

The column of thin blue smoke spiralled gently from the small, carefully built fire. Low flames from just a handful of deliberately arranged sticks danced beneath a tin billycan supported by two river stones. It was the manner of the man sitting on a low, flat granite boulder nearby, that the fire should be so. Sufficient heat to boil a cup or two of water does not need much wood; and the major took care that his presence in the unspoiled mountain habitat around him was unobtrusive.

He loved the mountains and had spent much of his working life in the Australian Army Survey Directorate, mapping the Great Dividing Range which runs the length of Australia's east coast. But the mountains also humbled him. He felt like a privileged visitor and moved with care lest the sounds of unseen life in the undergrowth, among the tangled branches of snowgums and in the glittering mountain streams, might suddenly cease and banish him to a sterile silence for the impertinence of his intrusion. At nights, wrapped in a sleeping bag on a groundsheet, hard against a fallen log to protect against the chill, whispering breeze, he would listen through the incantation of crickets for the soft pad of wallabies as they passed warily to drink from the nearby stream.

These were the times he liked best, when he could leave behind other men, who needed tents and conversation to protect them from the night. There were some who said Major Hugh Powell Gough Clews was of the mountains. He saw what others could never see and felt what others could only conjure with imaginative words. He breathed their air, scented by gums and alpine flowers, listened to and understood their voices and always took time to enjoy their beauty. This was where he knew he belonged.

On this morning in February 1950, the major—a title that clung even though he had ended his army days as a lieutenant colonel—should have been enjoying peace and solitude deep in the bush of the Blue Mountains to the west of Sydney. That was where he had decided to retire. He was sixty years old, but had been talked into a five-year contract as a senior surveyor for the construction of a big hydro-electric scheme in the Snowy Mountains—the highest ranges of the Australian Alps at the southern end of the Great Dividing Range. The Snowy Mountains are the only ranges in Australia high enough for lasting snow.

The plan developed by the commonwealth government was to halt the vast volumes of water from the spring snow melt that now wasted into the Tasman Sea—mainly down the fabled Snowy River—while regions to the west of the mountains suffered the miseries of drought. It involved collecting the melt in vast high-altitude reservoirs, then redirecting the water through tunnels under the mountains to the dry interior. On the fall from the high alpine dams, the water would also be used to generate hydro-electricity for the cities of Melbourne, Canberra and Sydney. It was an incredible proposal, unprecedented in Australian history, and would be one of the biggest engineering undertakings in the world.

The Snowy Mountains, however, were an uncharted wilderness. Little was known of their topography, geology or hydrology—crucial information which had to be gathered as quickly as possible before engineers could decide where to put access roads, let alone the intended massive dams, reservoirs, tunnels and power stations.

The man in charge of the surveying was Bert Eggeling, a subordinate of Major Clews in their army days, who had promptly hired the retired major when the project was

Surveyors crossing the Tumut River at Tumut Pond, September 1950.

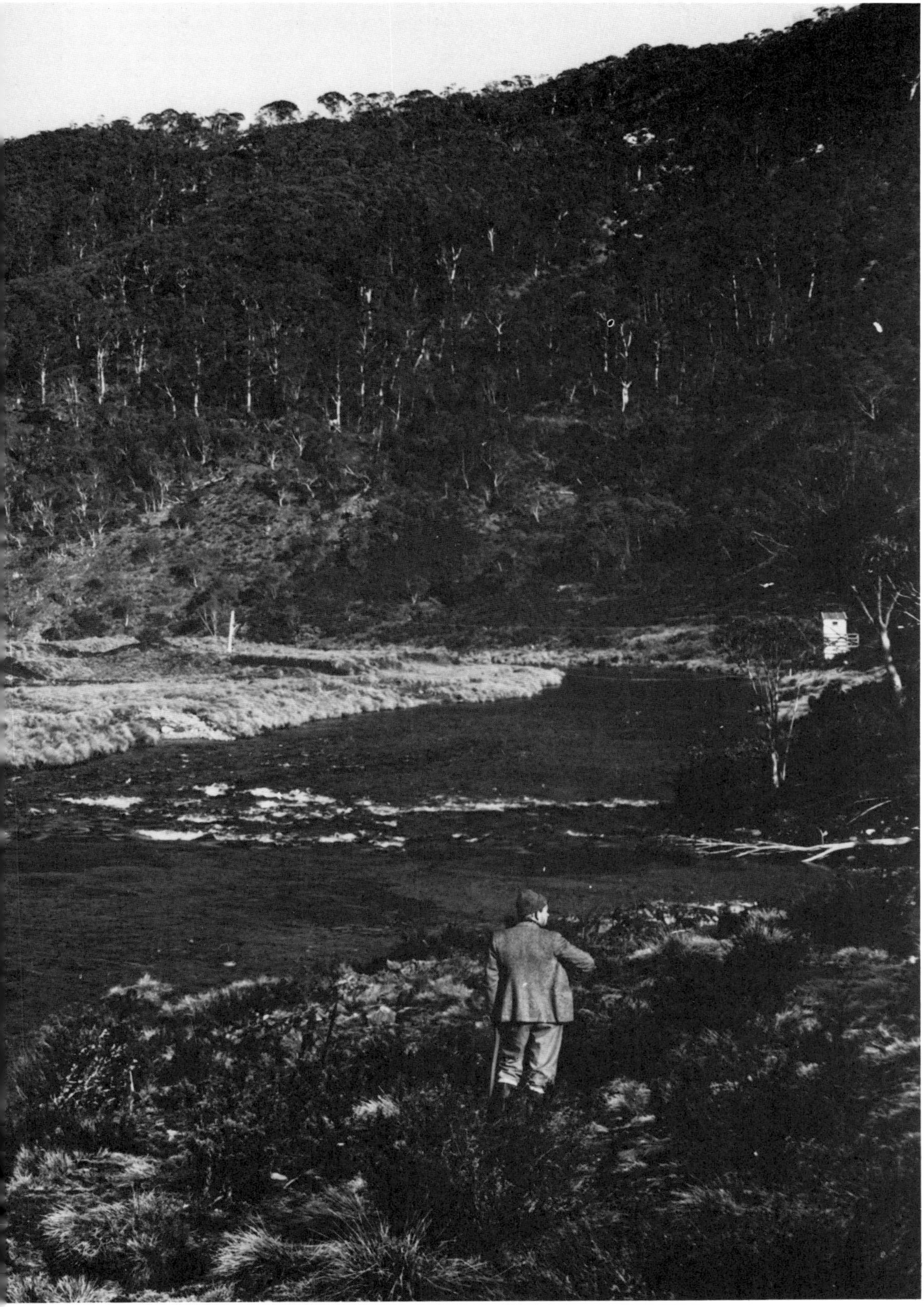

given the go-ahead. What little of the mountains had been mapped in the past had been done by the army and, in particular, Major Clews. So the major was back once more among the towering peaks rising sheer from seemingly bottomless gorges, the wild river torrents and almost impenetrable undergrowth which made movement slow and exhausting as well as dangerous. The density of the scrub and sudden mists could conceal precipices that dropped away for hundreds of metres.

Gazing out across the peaks and valleys in this craggy frontier, Major Clews often tried to imagine what it would look like with giant dams and roads, men and machines and even the townships that were planned. He would shake his head; it was just too much change to envisage easily. Once the giddy depths of the grandest valleys were filled with water, the wilderness would be conquered, the mountains tamed and forever subservient to the needs of humanity. Or would they? Perhaps he knew the mountains better than that. Either way, he trusted the scars would soon heal.

As he crouched by his breakfast fire, survey charts spread across the ground and weighted by small stones, Major Clews was aware that one age was ending and another beginning. The only people who had had close contact with these mountains before him and the teams of surveyors, geologists and hydrologists now clawing their way over the rugged terrain, were stockmen and, before them, Aborigines, who since time immemorial had followed the summer migration of the bogong moths from the plains into the mountains. Five generations of stockmen had herded cattle and sheep into the high country to graze during the summer months. They knew the mountains well enough but most confined their droving to stock routes established along the easiest gradients, where there was access to water. Few had need to explore the almost inaccessible areas into which the men of science were being sent.

Hydrologists measured stream and river flows, surveyors mapped the surface and geologists drilled deep into the earth's crust, delving for the secrets of the mountains' origins. The nature of the rock was crucial to the design and location of dams and tunnels. When the Scheme was announced, nothing was known of the region's geology and, though Snowy geologists learned enough for their purposes, their successors still debate the geological origins of the Australian Alps.

Examination of sedimentary rocks has revealed that up to 400 million years ago the Kosciusko region was a deep marine basin near the edge of a continent and prone to submarine volcanoes. Later, about 350 million years ago, the area had risen above the ocean and was adjacent to a shallow sea on the Canberra–Yass Shelf. The terrain was volcanic but the mountains were not formed by volcanoes. The two most common theories are that they were pushed up as fold mountains when Australia and New Zealand geologically separated, or uplifted by the reactivation of old fault lines which folded and cleaved the surrounding granite terrains. During the next 250 to 300 million years the region was gradually eroded, with the more resilient rock types forming the mountain ranges that have remained much the same for the past sixty million years.

Around sixty million years ago the Australian continent was much further south and a lot colder and wetter than it is today. Fossilised wood shows the Monaro Plain at the foot of the Alps to have been a substantial rainforest twenty to thirty million years ago. The forest disappeared with climatic changes and the continent's slow drift northwards—

Hydrologists measuring stream flows at Perisher Creek.

which is still occurring at the rate of between five and six centimetres a year. Volcanic activity continued in the Kiandra, Cabramurra and Tooma areas until seventeen million years ago, covering the Monaro with layers of basalt, which explains why much of the plain is treeless.

The first known animal life in the mountains is comparatively recent—just 30 000 years ago, around the start of the last Ice Age. About then, the Snowy Mountains became a haven for giant marsupials, including the giant kangaroo, which stood up to three metres tall, the Tasmanian devil and Tasmanian tiger. Humans first moved into the mountains about 20 000 years ago.

In Cloggs Cave, near the Buchan River in south-eastern Victoria, archaeologist Dr Josephine Flood has traced 17 000 years of almost constant occupancy, right up to the 1860s when the Aborigines in the region were killed off by measles and smallpox brought by the white settlers. In a wild gorge of the Snowy River north of Buchan are examples of Aboriginal cave art dating back 20 000 years—older than the famous paintings of the Lascaux bison hunters in southern France and Spain. In her book *Archaeology of the Dreamtime* Dr Flood has reconstructed the lives of the Cloggs Cave dwellers towards the end of the glacial period:

> During the daytime the rock shelter (at the cave entrance) was used; the rock ledges provided warm sitting places and a good vantage point over the valley. At night, fires were lit on the cave floor from Eucalyptus wood. The group gathered round, heating hearth stones and cooking food items collected during the day, whittling with scrapers to make wooden spears and boomerangs and rubbing hides with smooth river pebbles until they were pliable enough to sew together as cloaks. With sharp quartz flakes the possum or kangaroo skins were trimmed to size, holes were pierced with a bone point, its tip ground and polished to needle-like sharpness, and kangaroo tail sinews were chewed till supple enough to be used as thread.

The mountains took on a greater importance for Aborigines throughout south-eastern Australia at the end of the Ice Age when the climate had warmed sufficiently for bogong moths to discover that the coolest place to survive the summer months was in alpine crevices. Tribes from the coast, the western plains and what is now northern Victoria pursued the moth which was both a delicacy and a food for survival, particularly when the land below the mountains was seared by summer drought.

Archaeologists suggest that Aborigines were converging on the mountains for bogong moths as long as 8000 years ago—almost longer than the period of organised agriculture in Europe and Asia. It was an annual event in the natural cycle that governed the tribes' rituals. The profusion of the moth supply was able to sustain the large tribal gatherings that recurred for marriages, gifting ceremonies, initiations and corroborees.

The Snowy Mountains had a profound spiritual and material significance for Aborigines of the continent's south-east. By the time the Europeans arrived to clear the land for settlement, the Aborigines were among the last human beings still able to shelter in the spiritual anchorage of a pristine land. The name 'Bogong', bestowed on the peaks by the great tribes which populated the plains surrounding the Alps, was already old when the Achaeans were sacking Troy; it was ancient by the time Boadicea was attempting to drive the Romans from Britain.

In the timeless Aboriginal land, spring in the Alps was heralded not by calendars but by the radiant blooming of wildflowers. This started on the lower slopes with fields of black-eyed Susan, lilac hovea and purple sarsaparilla which appeared on the landscape like a colourful net over the humps caused by fallen trees and large boulders rounded smooth by eons of abrasion by Ice Age glaciers.

Even today, as the season progresses and the mountains warm, this many-hued profusion moves up the slopes until the summits and rolling alpine plains are seas of snow daisies, marigolds, everlastings, bluebells, heath and herbs. On the high plateaus sphagnum moss bogs protect primitive alpine plants. This wonderful flowering brings ravens and kestrels, which hover beneath the high, bright sun like ghostly apparitions in a glassy sea. Packed in the rocks below are the bogong moths.

The bogong moth, distinctive with its dark brown, sometimes blackish, colouring and dagger-shaped wing markings, is a nectar feeder. It breeds in the vast tablelands of southern Queensland and northern New South Wales and migrates to the mountains to undergo what entomologists call *diapause*—an adaptive mechanism which allows some insect species to survive unfavourable habitats or climatic conditions by subsisting for long periods on their own fatty reserves.

From November to early April bogong moths cluster in cool rock crevices. They do not mate or even move about. In 1865 an entomologist, Mr A. Scott, recorded an account of a visit to the diapause sites in the 'Bogong' Mountains by his colleague, Robert Vyner:

> The moths were found in vast assemblages sheltered within the deep fissures and between the huge masses of rocks, in which there form recesses and might almost be considered caves.
>
> On both sides of the chasms, the face of the stone was literally covered with these insects, packed closely side by side, overhead and under, presenting the dark surface of a scale-like pattern—each moth resting firmly by its feet on the rock, not on the back of others as in a swarm of bees.
>
> So numerous were the moths that six bushels of them could easily have been gathered by the party at this one peak... so abundant were the remains of the former occupants that a stick was thrust into the debris on the floor to a depth of four feet.

In more recent times, some migrating moths have been attracted to the vast concrete structures of Australia's new Parliament House in Canberra. The building had been in use for only a few weeks when millions of migrating moths, en route to the mountains, discovered its shaded courtyard walls and cool eaves and elected to shorten their journey.

When the summer heat is over, the moth returns to its original breeding grounds and the reproductive cycle begins once more.

Over thousands of years of summer trekking the Aborigines wove the moths and the mountains into their Dreaming, creating great ceremonies for the festival of the bogong moth. The tribes gathered first on the lower slopes, in places like Jindabyne and Blowering in the Tumut Valley, waiting for advance parties to signal that the moths had arrived.

Tribal elders then went up onto the higher slopes to perform sacred rites. When these rituals were completed a series of smoke fires, positioned down the slopes, indicated that it was time for the moth hunters to begin their sweep into the mountains.

According to custom only men collected the moths. They gathered them in 'nets' made of fibre from the pimelea shrub or kurrajong tree. When filled, the nets were carried down to camps on the lower slopes where women cooked the moths in prepared hollows of hot ash, or roasted them on flat rocks heated by hot fires of dry grass and leaves. Alternatively, they were ground into a paste and baked into patties. The moths' bodies were rich in fats, protein and sugar and early settlers noted that Aborigines returning from their summer in the mountains were fleshy and healthy-looking, their skin a glistening black.

At night, men danced around fires to a tempo set by women who beat drums of hide stretched over curled bark and held firm inside by stick ribs. As they beat the drums with nulla-nullas (clubs of wood or bone) the women chanted plaintive lullabies which drifted over the mountains and deep into the valleys. Some of them played reeds, using fingers and lips to extract an eerie buzzing sound.

But by the time of Major Clews the Aborigines were long gone. They had been annihilated within a few decades of the arrival of the first white settlers around the turn of the nineteenth century. Venereal disease, measles and smallpox as well as the European obsession with property and money—or, more directly, the weaponry and legal system used to protect them—all contributed to the Aborigines' demise. A nomadic people who believed the land belonged to all had no defence against a new order so alien to everything they had known for tens of thousands of years.

In the Alps, as elsewhere, the Aborigines made no significant environmental modifications. Unlike the European settlers who replaced them, they accepted nature's authority.

The Snowy Mountains are like no others on Earth, and not only because of the moths and the spells of an ancient Dreamtime. Nowhere else do eucalypts grow in the snow, with the genus clearly marking the snow line—above it, snow gums, stripped and twisted by blizzards that strike with terrifying ferocity; and below, tall straight woolly butts, elegant mountain ash, peppermint forests and towering ribbon gums, their shredded bark hanging in long sheets like dried vine curtains in a desiccated rainforest.

More than a century after the Aborigines' disappearance from the region, Major Clews, alone in a grove of peppermints and gums high on the mountain slope, was comfortable with the spirits of the past which moved in the untouched bush that enveloped him. He idly prodded his fire with a stick, savouring the sweet eucalyptus smoke and the warmth in his mug of black tea. His simple breakfast consisted of cold meat and cheese sandwiches, thrown together before he left his main camp at Dry Dam, already a full day's march away.

There were no roads and the major disliked riding horses; so he walked. Scattered throughout the mountains were small teams of scientists and workmen looking for the best sites for dams, power stations and roads. The major met them all as he traversed the mountains, organising his survey parties. He commiserated with the drillers who had to manhandle their heavy drilling rigs, piece by piece, over the mountains. It was a hellish task and he was glad it wasn't his lot. Three or four days often passed before Major Clews returned from his excursions to base camp, a world where industrious men with noisy jeeps and two-way radios talked excitedly as they dealt with orders and queries from headquarters in Sydney, nearly 500 kilometres away.

The morning's first bird calls cracked the still air: the haunting whiplash cry of currawongs, the chatter of multicoloured parrots and the occasional screech of a black cockatoo. Cockatoos are nature's messenger in the Australian Alps; they know of an approaching storm long before any man. Flocks warn of an impending blizzard, filling the air with ear-splitting shrieks as they wing to safety on lower slopes.

The major stood bare-chested in the crisp morning air, his mug and billy now rinsed and dripping on black granite by the edge of the stream. He carefully wrapped a rubber pouch, high and tight, around his chest. Inside were survey maps and mail for the men he would meet later in the day. Almost all the mail carried against the chest of this wiry, ageless man was posted a world away—in unpronounceable villages and cities scattered across the breadth of Europe. He replaced the shirt, tucking it into baggy serge trousers held at the waist with a simple, handmade leather belt. The legs were tucked into thick woollen socks that bulged above heavy, rubber-soled army GP boots.

After repacking a small rucksack with his bedding, billycan, mug, tea, sugar, pipe tobacco and the flask of rum he carried everywhere for a nightly tot, the major carefully smothered the embers of his small fire. Taking a last indulgent look around, he planted his cherished 'mounties' hat on his head and strode purposefully into the bush. With early morning sun on his back and mountain air in his lungs, a man felt good to be alive.

The major's long, lone marches through the mountains were already becoming legendary. Kon Martynow, a 27-year-old former forestry student, had experienced the major's idiosyncrasies at first hand. A Russian-born Pole, who as a youth had wanted to be a forester, Kon had spent the previous five years in a displaced persons camp in Germany. Before that he had endured another camp—a German slave labour camp. He had been in the mountains just a few days when the major decided to introduce him and Edward Wysocky, a Polish surveyor, to the bush, and to the tasks ahead of them.

Kon and Edward had been in Australia for only three weeks, and memories of the soft European woodlands were still fresh in their minds. They were unprepared for the dense, hard scrub and the remoteness of the Australian bush.

The major had decided to take his two new charges on a reconnaissance of the site chosen for the Tumut Pond dam. The engineers had selected a location on the river. Now the surveyors had to run traverse lines, measuring horizontal and vertical angles to profile the fall of the river and adjacent terrain, in preparation for the mammoth construction and tunnelling operations. After hours of fighting their way through dense undergrowth, Kon and Edward emerged on the edge of a sheer drop of more than 300 metres to the Tumut River, which raged below them in an emerald and white torrent over lethal-looking rocks. Sweat ran from their every pore and stung the exposed skin of their faces and arms scratched raw by the clawing, twiggy undergrowth.

The two stood nervously, waiting for the major; but after several minutes he had not appeared. They waited, hoping the major would suddenly appear behind them to lead them safely to the river. The Australian bush was a daunting enough experience, without the prospect of having to descend the face of a mountain. They tried calling, 'Major...Major'; it was one of the few English words they knew. Eventually they convinced themselves the major must indeed have gone by another route. Feeling it was their fault for losing him, their guide and boss, they backed cautiously onto all fours and edged over the precipice to begin their descent.

It was dangerous. A single slip and they and their dreams of a new life would be smashed on the glistening rocks below. But neither man felt he had much choice. Venturing back into the bush presented, in their minds, only a marginally better prospect. At least this way they felt comforted by the thought they would be reunited with the major.

As they inched their way down, they grasped bushes for support. Suddenly, one came away in Kon's hands. 'I didn't see any rock crevices, but my fingers found them very quickly,' he later told his friends. After what seemed an age—every muscle tensed, every toehold carefully measured—the two men touched firm ground. Looking up they could scarcely believe they had made it safely. The rock face seemed to rise sheer to the azure sky. They called to the major but the roaring river beside them swallowed their cries.

After two uneasy hours, there was still no sign of Major Clews. The afternoon was advancing and neither man wanted to spend the night by a raging torrent at the bottom of this hostile gorge. Somewhat nervous about what the major would say when finally they met again, they began picking their way downstream, following the course of the river until the mountain wall which hemmed them in became less severe. On an easier slope they climbed dejectedly out from the gorge's deepening gloom. By the time they reached the top it was dark. Although they had no food, they had little alternative to camping for the night in this alien world with its eerie nocturnal sounds.

Despite the heat of the day, the night chilled quickly. The men lit a small fire but neither slept. They spent most of the night tossing from side to side, seeking warmth from the fire's low flames. As one side of the body heated up the other nearly froze. Roll over, thaw and heat; roll over thaw and heat—the night dragged on in an eternity of discomfort.

The next day, tired, hungry and dispirited, the two men pushed on towards where they believed the campsite lay. They found it some hours later and congratulated themselves on their navigation. But the camp was empty. All that greeted them was a note from Major Clews which they painstakingly translated: 'Sorry lost you—but you know now where the dam site is, so off you go and do it.'

While the perplexed pair were gathering together the instruments they would need, three labourers strode into camp, sent by the major to cut a bridle track through the scrub to the river—in a wide gentle arc around the fearsome gorge.

About three months later Kon's memory of this episode was triggered by a glimpse of the major backing down a similarly sheer slope high on the bank opposite to where he was working. Kon chuckled to himself as he focused his gaze on the steep face. Bushes again parted momentarily to reveal the outline of the old man inching his way down the almost vertical descent.

When Major Clews had left his overnight camp, the day had got off to a promising start, but then had turned bleak and drizzly. Kon, hunched against the grey, chilled air, was peering intently through a theodolite when the movement on the slope first caught his attention. He stopped to watch the major's progress. The man was amazing. A tough bush character, he was old enough to be their father—indeed, in many ways he *was* a father to them. And now here he was again, appearing, it seemed, from nowhere to see how they were getting on.

Kon was eager to make him welcome. 'Oi,' he yelled, beckoning to Edward Wysocky, who was in charge of the team, and two chainmen, Stan Kajpust, a fellow Pole, and Albert Kalvis, a Lithuanian. Kon lifted his wrist and tapped the face of his watch, strapped to an arm that the Australian sun had already turned a dark leathery brown. The men waved acknowledgement. The roar of the falling, rushing water drowned the noise of their activity at the bottom of a deep gorge that flanked a bend in the Tumut River.

The team was working methodically along the base of a granite wall gouged by raging torrents. Two were working with theodolite and pole; the other pair was hacking a rough path where sheer granite slopes gave way to a thicket of dense scrub. From high on the opposite ridge, where the major had begun his descent, they were ant-like figures, dwarfed by scrub-covered mountain tops that towered hundreds of metres above the narrow rocky beach.

Kon left the theodolite and walked to where a cluster of black rocks formed a low breakwater at the river's edge. In the shallows behind the rocks was a large cast-iron cooking pot, its curved top and handle protruding just above the water's ruffled surface. The river acted as a natural refrigerator. The men cooked every two or three days—salted beef with potatoes and carrots—then stored the food in the river. Kon lifted the pot from its watery closet and began climbing the slope to the campsite they had established above the river. From the top of the rise he paused to look back. On the opposite bank Major Clews had reached the bottom of the cliff and was carefully choosing the best place to cross. Not for the first time, Kon shook his head. The water would be waist deep and icy.

Lifting his gaze above the distant, tiny figure of the major, Kon swept his eyes slowly across the magnificent vista. The Australian bush was a wonderment. Every single tree and bush was new; nothing was the same as in his homeland. No scent here of birch or pine resin. The perfume of eucalypts predominated—almost intoxicating if one inhaled deeply.

This ancient continent, which predated, in geological terms, almost every other landmass on Earth, had a unique beauty and such strange birds and animals—the platypus, like a cross between a big water rat and a broad-billed duck; the shy, furry koala that curled into the forks of branches high above the ground; the lyrebird with its enchanting call and mating dance; and the comic, stolid wombat, that rooted for food in the thick undergrowth. But most beautiful of them all was the grey kangaroo, graceful like the deer, but unlike any other animal on Earth. Regrettably, most parts of the mountains were too steep for kangaroos, so they were a rare sight.

By the time the major arrived, pink-cheeked and grinning broadly, all the men were in the camp and ready to eat. 'Hullo, hullo,' he nodded to each man. They grinned and nodded in return. Major Clews was a welcome visitor, even if language differences limited conversation. The surveyors could communicate in technical terms and with frequent tapping on maps and charts, but the manual workers, the chainmen, had to be content to reside for the time being in their own world.

As one of the chainmen sliced cold meat and dished out vegetables, the major pulled off his boots and socks to dry by the fire. The men watched eagerly as he removed his shirt, and then the rubber pouch. Tantalisingly, he withdrew tobacco and filled his pipe. Then, with a grin, he delved into the pouch for the precious mail. The letters were taken quickly to the security and privacy of the tents where later that night they were read and reread in the yellow glow of kerosene lamps. From mothers, wives and sweethearts, anxious and eager for news and full of chatter about lives in another world, the letters briefly transported young men home.

During the meal the men joked with each other; mateship transcended language barriers. The major joined in and when he laughed, his dentures rattled noisily, sending his listeners into hysterics. He stayed just an hour—long enough to dry his footwear, have a smoke, eat his meal and savour a cup of hot tea. It was long enough, too, for a perceptive man to monitor the men's progress and to measure how they were coping with the isolation and primitive conditions.

Most newcomers, both migrants and Australians, were from towns and cities. Some climbed aboard the first truck back to civilisation after just a day at base camp. To survive the rigours of the mountains a man needed to love the bush, to find within it peace that outweighed such torments as loneliness, ants and bush flies. When sweat and campfire smoke matted his hair and dried his tongue he needed to be able to savour, not shrink from, the cold river water.

Under the watchful eye of the major, a strict environmentalist who forbade even the killing of snakes, most of his foreign charges became true bushmen, taking on a character that most native-born Australians had long consigned to folklore. The major liked these wiry Europeans—men with deep eyes and hard bodies who had been toughened by war and deprivation. They had been to hell and back but had arrived, finally, with their souls intact, in the unspoiled Australian bush.

His visit over, Major Clews repacked his pipe and tobacco in the rubber pouch, along with mail which he would put aboard the next truck to Cooma. By the time the men were returning to their work at the river's edge, the major was on his way back to Dry Dam, the summer base camp high up on the ridges. There, amidst the dust, jeeps, canvas and stout wooden tent poles, a semblance of military order prevailed. But conditions were still primitive and would remain so until tents were replaced by wooden huts and barracks and the campsite gave way to the township of Cabramurra.

At night the oil lamp in Major Clews's tent was always the last to die. His day began at dawn and did not end until he had carefully checked survey figures and logged the day's work in his field book. The survey sheets then had to be bundled together to be placed in the following day's mail bag to Cooma, from where they were sent to the Snowy Mountain Authority's head office in Sydney.

The surveyors and geologists worked under great pressure. Construction could not begin without the scientific data they collected, but there were political pressures to achieve tangible results as soon as possible. The task was enormous. The geography of damsites had to be analysed in exhaustive detail before dams could be constructed; the composition

Surveyors crossing the Swampy Plains River in the early 1950s.

of granite, thousands of metres beneath the mountains, had to be known before tunnels could be drilled; the terrain and size of the Scheme called for new technologies and a mammoth work effort. For the sake of speed, the tunnels were to be driven in from each side of the ranges and were to meet in the middle. To achieve this, and to avoid the worldwide ridicule that failure would have brought, the surveys had to be flawless.

When the bookwork was finished, Major Clews liked to relax with a tot of rum, listening to the wind as it lashed branches and punched canvas, and to the distant roar of a plunging river. He thought about the work, the future of the Scheme and how it might change Australia. He thought of the men, out there in the bush, full of dreams and hopes for a new life.

Sometimes he gazed into his own future. He wished only for a simple life—somewhere peaceful and in the mountains, where he could indulge his great love for breeding flowers, particularly dwarf dahlias and roses.

CHAPTER TWO

The Tumut River Gorge showing the Happy Valley camp built for workers blasting the access tunnel to Tumut-1 power station.

POLITICS OF WATER

The devastation in Europe at the end of World War II was unprecedented in human history. Great cities had been turned to rubble. Millions of people were homeless, starving and facing a bleak future.

The end of the war had brought great relief but left people, particularly the losers, in an uncertain world. History has meticulously documented the paths and fates of the politicians, generals and the more visible heroes. But time rapidly washed over the ordinary soldier—stripped of his uniform and forgotten, an anonymous statistic among the war's human flotsam. Armies dissolve as rapidly as they appear. But as with all war, men who had lived on a razor's edge for so long found that returning to an expected normality produced hidden horrors as mind-breaking as combat.

What faced the Luftwaffe pilot after he returned home, the U-boat captain, the Italian tank commander, the Polish airman who had been fighting from exile in England, the resistance fighters of all invaded countries and the battle-weary Englishman who found victory to be a double-edged sword? Many found normality impossible to achieve. Their loves, dreams and hopes had been crushed under the rubble of their own making.

Even at the war's close the dying and killing took time to end. The victors were taking revenge in the courts and new powers in Eastern Europe were using the confusion and loss of public records to expunge opposition. In a beaten Germany, law and order was slow in returning. Black market operators waged a deadly entrepreneurial war, while unexploded shells and booby traps left by armies and fanatics continued to kill and maim children and drifters lulled into thinking the fields and forests were suddenly safe.

But an even greater problem that faced both military and civil authorities were the tens of thousands of survivors of Nazi concentration and slave labour camps, as well as the several million displaced persons and refugees who were left in the wake of the Soviet takeover of Eastern Europe. The Allied authorities had to find homes for more than eight million displaced people. Eventually, about six million were repatriated to homelands in Europe and to the new state of Israel. But more than two million did not want to go back, or had no homeland to return to. Many came from countries which were in the process of being erased from the global map.

On the other side of the world, a passionate plea was made to a young developing country beyond reach of the devastation, to offer a home for these people. Time was pressing. The United Nations Relief and Rehabilitation Administration went direct to the Australian Prime Minister, Ben Chifley, and asked him if he would accept 100 000 homeless Europeans.

Chifley almost choked on his pipe when he heard the figure—it would be a big injection of 'foreigners' into a relatively small and homogeneous Anglo-Saxon population. He promised the United Nations representative, Sir Robert Jackson, that he would do his best to convince Cabinet of the need and of Australia's responsibility. Chifley used the ploy of telling his Cabinet colleagues that if Australia moved quickly, it would get the pick of the bunch. It was a clever strategy, and decades later Australia is still profiting from the skills and enterprise introduced by so many of those who came.

The request from the United Nations coincided with plans by the Chifley government to broaden the base of the Australian economy by hastening the development of secondary industry. It also came as the government was attempting to grapple, once and for all,

Prime Minister Ben Chifley.
PHOTO: NATIONAL LIBRARY CANBERRA

with the continent's critical and often crippling lack of water. It was thrashing out the final details of one of the boldest engineering projects ever conceived—the harnessing of melted snow on the Australian Alps through a vast irrigation/hydro-electric scheme. Such a scheme, as well as the postwar development projects that the states were clamouring for, meant the country was going to need a pool of labour far beyond the capacity of its existing population.

So Australia said 'yes', and tens of thousands of people from Europe's razed cities and refugee camps began a journey that would transform not only their lives but the face of an entire nation. With them came many of the war's other victims—the discarded combatants of both sides. They were sent into the mountains—enemy and ally, oppressor and victim—to work together.

They entered the isolated realm of the Australian high country stockman, the reticent, resolute character embodied in Australia's most cherished legends. It was the strangest mixing of cultures since the construction of the pyramids. Yet together they built an engineering wonder of the modern world.

The Snowy Mountains Hydro-Electric Scheme, as it was christened, had been a long time coming. It is a tragic irony that it was made possible only by the devastation of war and the creation of an instant, and politically convenient, workforce.

Lack of water is a constant problem for Australia, the world's driest land. Along the eastern side of the continent, most of the water run-off from the Great Dividing Range flows eastwards into the Pacific Ocean and Tasman Sea, leaving the vast tracts of land on the western side prone to lasting drought. Rivers that flow naturally westwards are generally meandering and slow.

Evaporation in the hinterland is high and in dry seasons many tributaries of the main inland rivers—the Murray, Murrumbidgee, Lachlan and Darling—become little more than chains of waterholes. The Snowy Mountains Scheme would dam east-flowing rivers to create huge man-made alpine lakes, from which water would be diverted westwards in massive tunnels cut through the mountains. The plan was for the water not only to irrigate the parched interior but to generate hydro-electricity for cities of the eastern and south-eastern seaboard.

As far back as the early 1800s men of vision had begun to see the potential of the Alps' melting spring snow. At their highest point, the Snowy Mountains give rise to a number of rivers, the biggest being the Murray, Murrumbidgee, Tumut, Tooma and Geehi. The last four all dump into the Murray which flows westwards to reach the sea 2500 kilometres away in the south-eastern corner of South Australia. Yet despite its length, for most of the year the Murray's wide bed carried just a trickle. Before the construction of the Scheme the bulk of the vast alpine run-off flowed eastwards into the Snowy River, turning it, after the spring thaw, into an unchecked torrent that raged along an erratic course to pour into the sea off south-eastern Victoria. In peak flow, the Snowy wasted into the Tasman Sea at the rate of two million litres a minute.

It was a tormenting waste when the rest of the continent suffered a chronic lack of water. If Australia's total annual run-off were to be spread evenly over the continent, the water depth would be a mere three centimetres.

Time and again lack of water had almost put an end to the settlement of eastern

Australia. Between 1813 and 1815 drought and the spectre of famine added urgency to efforts to find a way from the Sydney area over the Blue Mountains. Fortunately for the colony, explorers Gregory Blaxland, William Lawson and William Charles Wentworth finally found a way through the mountain barrier and discovered seemingly well-watered plains. But even this pastoral vision was temporary.

Widespread drought struck again from 1824 to 1829, withering the newly discovered plains country. Crops burned, stock died and, despite their familiarity with the land's natural resources, many Aborigines were also reported to have perished through starvation.

For the settlers, the most severe impact was on the flocks of sheep they were trying to build up. By the 1830s the sheep industry was well established in every Australian settlement. Through drought and fluctuating world markets it rode a wild see-saw of booms and busts. In one respite from drought in the early 1830s, which coincided with buoyant markets, Governor Gipps of Victoria reported the country around Melbourne 'strewn for miles, almost hundreds of miles, with champagne bottles'. Men, almost on their knees, suddenly found themselves rich beyond their wildest dreams and the fingers of exploration spread further as they sought new lands. Yet hundreds were just as quickly ruined by a single dry season. The familiar visitation of disaster occurred again from 1837 to 1840 and even the snow-fed Murrumbidgee was dry in places, allowing settlers in some districts to run horse races on its bed.

In March 1840 the Macarthur–Strzelecki expedition discovered more pastures and named Australia's highest mountain. The new pastures were welcome and the christening of Mount Kosciusko was an historic milestone, but neither achievement offered respite from the cruel climate.

In 1843 good sheep were selling for three pence a head. The list of bankruptcies grew longer daily. The Bank of Australia failed. James Tyson, the king of the squatters, sold one run for a tot of rum and a second for twelve pounds, which he never bothered to collect. The infant sheep industry resorted to the desperate expedient of boiling down sheep for tallow. The average sheep yielded twelve to fifteen pounds of tallow and tallow was worth five cents a pound—at least for a while. By the end of 1844 more than 200 000 sheep had been boiled down and tallow prices began to fall.

Drought struck again in the late forties and for more than a decade from 1861 to 1870. Some graziers built dams on creeks and guarded them with armed men against reprisals from aggrieved landowners further downstream. In a dry season in 1858 parties of men destroyed or damaged more than twenty-five dams on the Yanko Creek, which runs between the Murrumbidgee and Murray rivers. The story is told of a pastoralist who consulted a Sydney barrister when he heard that a group armed with firearms and spades intended cutting a dam he had built on the lower Lachlan River, which joins the Murrumbidgee. The learned counsel advised his client simply to be ready with a bigger and better mob of armed men.

Still the droughts continued, bringing pain and despair. A report from Cowra in New South Wales in January 1866 described scorching winds, raging bushfires and flocks of parrots and other birds devouring what wheat crops remained after many rainless months. Rainless seasons, which continued to strike randomly through to 1878, almost wiped out whole pockets of the sheep and wool industry. Meanwhile, great volumes of water from the Alps were running off into the sea. Men looked longingly at the potential for harnessing these but were still a long way from putting their bold ideas into practice.

In 1884 the New South Wales surveyor-general, Mr P.F. Adams, told a Royal Commission on the Conservation of Water that any permanent system of irrigation in the colony would have to be drawn from snow waters. Adams proposed that water be diverted from the Snowy River at a point about eight kilometres above the junction of the Snowy and Eucumbene rivers (near today's Island Bend Dam). He concluded that by constructing a canal across a gap in the Great Dividing Range the water could be channelled into the Murrumbidgee and used to irrigate vast inland areas.

Adams made this report more than a century ago, long before the tools and techniques were available to construct roads, tunnels and dams in mountains. But he was the first to hold aloft the key which a future generation would use to unlock the area's potential.

In 1888 New South Wales recorded its driest year ever. An estimated one million sheep passed through the town of Wagga Wagga in search of grass and water. Stock routes along the north bank of the Murrumbidgee were closed. By 1903, settlers in New South Wales, Victoria and South Australia had been battling the effects of drought for more than half a century. It is a wonder any still had the spirit to keep going. During the drought period from 1895 to 1903 Australia's sheep population halved, from about 106 million to about fifty-three million. It took thirty years to make good this loss.

The realisation that it was necessary to make better use of water resources was an early driving force behind the push to weld the colonies into a nation. But even after federation a half-century of argument and wrangling was to elapse before the states with the strongest claim to water from the Alps could agree to a plan. New South Wales wanted to divert the upper reaches of the Snowy River into the Murrumbidgee, solely for irrigation. Victoria wanted the water concentrated into the Murray River—initially to allow reliable river transport and irrigation, and later for hydro-electricity and irrigation.

One of the main sources of inter-colony and, later, inter-state bickering was the fact that the Murray River forms the border between New South Wales and Victoria. South Australia, as the final recipient of the river, also had a vested interest in any moves to exploit its resources.

The debate raged through a string of inquiries and royal commissions, perpetually divided on whether the water should be used for irrigation in a Snowy–Murrumbidgee scheme, or, as time went on, for hydro-electric development in the Snowy–Murray scheme. The balance began to tilt towards hydro-electricity during World War II, when the commonwealth government became concerned about the vulnerability of its coastal thermal power stations.

The Commonwealth's involvement injected both an urgency and a degree of objectivity into the long-running saga. In 1944, for the first time, a dual-purpose scheme was proposed. A Commonwealth and States Snowy River Committee was set up to investigate the proposal in detail. It was pushed vigorously by the Prime Minister, Ben Chifley, and his Minister for Works, Nelson Lemmon, a wheat farmer who represented the Western Australian electorate of Forrest.

Lemmon's enthusiasm for the scheme reflected the enhanced sense of nationhood that had been forged by Australia's participation in two world wars. Lemmon came from

Nelson Lemmon.
PHOTO: NATIONAL LIBRARY CANBERRA

Ongerup, a small farming community on the other side of the continent. He represented a state and locality that would gain no direct benefit from the scheme.

The New South Wales government, however, continued to push irrigation as the main object of the scheme. It also argued that electricity generated by the Snowy would be no cheaper than that produced by its coal-fired thermal stations. An important factor behind the state's stance was that the New South Wales Electricity Commission saw the Snowy scheme as a rival. It had big expansion plans for its coal-fired thermal power stations and saw the Commonwealth's proposal as a threat to these ambitions. Another reason the state fought so stubbornly for the irrigation priority was the seriousness with which it viewed the effect of drought on sheep and wool production. This was, in some way, explained a decade later by Dr M.C. Franklin, a distinguished drought expert. He calculated that the drop in wool production and sheep numbers between 1944 and 1946 cost the Australian economy more than one billion dollars over the succeeding decade.

But the Commonwealth was equally determined and, in a political move to try to blunt the New South Wales attack, went so far as to say it doubted the value of irrigation in preventing drought losses—at least among sheep flocks. It pointed out that sheep numbers in the Riverina district had declined from 3.1 million in 1940 to 1.6. million in 1946, yet this area contained Australia's largest single irrigation project. It deduced that if losses of this magnitude could occur during drought in a region surrounding an irrigation area, then irrigation had little effect on pastoral drought. Rather, it argued the highest returns from irrigation were obtained when farmers used the water to produce rice, dairy or horticultural products—and markets for those commodities at the time were saturated.

The Commonwealth insisted, therefore, that power should be the basis of any scheme to utilise the Snowy waters, and that irrigation should be a by-product which would expand agriculture away from its focus on grazing. It argued that the economics of the scheme should be based solely on its power-producing potential, but that water discharged from the power stations should be supplied free for irrigation.

Another plank in the Commonwealth's platform was that the scheme would lie geographically midway between Sydney and Melbourne and thus serve both cities—which had rapidly increasing demands for electricity—equally. It also noted that hydro-electric power would conserve the equivalent of one and a quarter million tonnes of coal a year. As a final measure, the Commonwealth used national defence as its trump card against the New South Wales demand for 'irrigation-only': it was essential to provide power generation in areas secure from enemy attack.

The Commonwealth's dual-purpose proposal was based on finding a means of storing great quantities of water at an altitude that would permit a controlled release into the Tumut River, and thence to the Murrumbidgee and Murray rivers. It also depended on being able to use the water for electricity generation as it fell to the lower levels before passing to the irrigation areas.

The final plan, presented in November 1948, consisted of two physically separate projects. The northern project would divert water from the Eucumbene, upper Murrumbidgee and upper Tooma rivers into the Tumut River. These waters would be used to generate electricity in the Tumut Valley during their swift fall to the plains. They would then flow, via the Tumut River, into the Murrumbidgee for irrigation. The main storage for this system would be a reservoir formed by damming the Eucumbene River near Adaminaby. In the southern project, water would be drawn from the valley of the Snowy River, diverted into the Murray and used to generate power in the course of its fall. The storage for this would be a reservoir created by damming the Snowy River at the bottom end of the Jindabyne Valley.

New South Wales, however, still held out for its own scheme, forcing the Commonwealth to invoke its defence powers and put through legislation giving it total control of the alpine headwaters and the development of the Scheme. The Snowy Mountains Hydro-Electric Power Act, operative from 7 July 1949, also encompassed the establishment of a Snowy Mountains Hydro-Electric Authority to construct and operate the Scheme.

Earlier, in a broadcast 'Report to the Nation' in May 1949, the Prime Minister, Mr

Chifley, declared:

> The Snowy Mountains plan is the greatest single project in our history. It is a plan for the whole nation, belonging to no one State nor to any group or section. It is a two-sided plan, because it provides not only for the provision of vast supplies of new power but also for an immense decentralisation of industry and population. This is a plan for the nation and it needs the nation to back it. I trust that you will all keep yourselves informed of its progress. I recommend that you listen to the discussions on it when they take place in Parliament.

It was believed that the Scheme, particularly the irrigation, would lead to the creation of two large inland cities in the Murray Valley and Murrumbidgee Valley, each with a population of about a million people.

But there was strong resistance from the opposition—a Liberal Party–National Party coalition led by Robert Gordon (later Sir Robert) Menzies. Menzies attacked the Chifley government for brushing aside the states and for assuming a power which he claimed it did not possess: and for enacting legislation therefore tainted with serious constitutional illegalities. That aside, he admitted that the proposed scheme was 'bold, comprehensive and well designed'.

This question of constitutional validity was to trouble the Snowy Mountains Hydro-Electric Authority for almost a decade until the New South Wales, Victorian and South Australian governments finally agreed to validate the scheme in their state parliaments. Until then the legal floorboards of the Scheme were described as 'creaking ominously'.

Three Germans who had been recruited for the Scheme sought to leave before their two-year contract had expired. The Authority started legal action against them to recover the outstanding balance of the fare to Australia. Their solicitor threatened to apply to the High Court for a determination on the validity of the Act under which the Authority was constituted. The Authority immediately received instructions from Canberra ordering it to drop its action and thus avoid the court challenge. The Australian Workers Union also used the threat of a High Court challenge to the Authority's legal standing to force it to allow the workforce to operate under state industrial awards, which were far more favourable to workers than federal awards.

The construction of the Snowy Mountains Scheme required an unprecedented engineering feat and a workforce that Australia could not possibly provide. As well, the commonwealth government had promised the states it would not draw off labour they needed for postwar redevelopment. Thus was precipitated one of the greatest experiments ever attempted in mass migration. It reshaped a young Anglo-Saxon/Celtic country into a nation of diverse nationalities. It sowed the seeds of a free, multicultural society based on ethnic tolerance.

The war and its atrocities were still sharply etched into the minds of the young men who flocked to join the Snowy workforce. But in the primitive workcamps high in the Australian Alps Englishmen, Germans, Italians, Austrians, Poles, Greeks, Dutchmen, Portuguese, Spaniards, Hungarians, Swiss, Swedes, Finns, Czechs, Lebanese, Latvians, Russians, Danes, Cypriots, Ukrainians, Americans, Turks, Frenchmen and Norwegians—

Prime Minister Sir Robert Menzies.
PHOTO: NATIONAL LIBRARY CANBERRA

more than thirty-three nationalities in all—shared hard work and laughter, ate from the same cooking pots, drank at the same bars and vowed to keep ethnic hatreds out of this young country which promised them all a new life.

The one exception was the antagonism between Serbs and Croatians. These were the only ethnic groups to bring their hatred to Australia. When the Cooma Shire Council erected poles for the national flags of the people who built the Scheme, it was forced to retrieve the Yugoslav flag each night. It had learned from experience that it would otherwise be gone by morning. On one occasion a group of children watched as a Yugoslav royalist, whom they mistook for a council worker, sawed through the entire flagpole.

But for others the mountains provided a release, a chance to smash into something, carve out a new future and wash themselves in sweat to cleanse the nightmare. For many it provided their first sense of peace:

A burst of light marked the end of the first reel. Workers in a mess hall in Cabramurra were watching an early Hollywood recreation of the British Army's evacuation of Dunkirk. As the projectionist threaded the second reel, a heavily accented voice, unmistakably German, rose above the babble: 'What a load of rubbish...it was nothing like this.'

A Cockney voice, coloured with indignation, responded: 'How would you know?'

'I was there,' said the German.

The Cockney didn't believe him.

'I was in the air...flying a Messerschmitt,' the German explained.

'...Christ, well I was there too, mate. Trying to get into a bloody boat.'

The two scraped their chairs together and learned that their opposite roles in the French seaside hell were played out on the same stretch of sand and water.

It transpired that a broken jetty the Englishman used to reach his boat was destroyed soon after by the fellow now sitting beside him in a draughty mess hall in Australia's Snowy Mountains. The two men were seen some hours later leaving the wet canteen, an arm over each other's shoulder, on the way to becoming life-long mates.

The workers remembered, and were often reminded of, the words of the Scheme's first commissioner, William (later Sir William) Hudson, who had toured the displaced persons' camps offering work 24 000 kilometres away: 'You won't be Balts or Slavs...you will be men of the Snowy,' he had told them.

For perhaps more than 200 men the dream was never realised. The death toll during construction of the Scheme was horrendous. The official figure of 121 is only an estimate. One policeman has stated that he alone handled about 130 inquests and he was only one of many policemen based in the mountains during the construction. The Authority's figure is based on details it obtained two decades later from its own personnel records and the remaining files of contractors. As well, the kinds of accidents that occurred often made identification difficult.

The work was dangerous and the environment often hostile. High explosives, sheer mountain slopes, perilous roads and language barriers contributed to a grisly accident toll. Many of those killed suffered agonising deaths, stretched out in the remorseless summer heat or lying in frozen winter mud waiting for help which all too often simply could not get to them. To some, they were 'industrial heroes'. Others saw their acceptance of the risks in the quest for a better future as an indication that the Australian pioneering spirit did not die with the nineteenth century.

At the unveiling of a memorial at Cooma in 1981 to the men killed on the Scheme, an elderly Spaniard stood quietly watching the ceremony. Like a fighter pilot who records kills on his fuselage, he was wearing a construction worker's hat adorned with the names and dates of places he had worked during twenty-three years on the Scheme. He had returned to Cooma to pay tribute to four mates—a Frenchman and three Italians—who were killed when a winch broke and sent three of them plunging to their deaths. The fourth was killed when the snapped cable followed him to the bottom and sliced him to pieces.

Many fatalities were caused by the steep, rough tracks which often iced over. Driving a truck in the Snowy required the nerves of a tank or U-boat commander. Access tracks to the construction sites of dam walls and power stations were so steep that cement trucks had to be lowered on a steel cable from a bulldozer. Sometimes the cable broke and the truck became a crumpled wreck far below on the river bed. Drivers soon learned to leave their trucks during such an operation.

Drivers risked their jobs if they abandoned their vehicles simply because they were frightened. But after several losses on the perilous mountain access roads, the rules were eased slightly to allow the driver of a truck that was beginning to slide to leave his seat and steer from the running board in case he needed to jump free. Sometimes drivers dragged logs lengthways behind them as brakes. In winter it was standard practice to steer a sliding vehicle into a snow bank and hope it was deep enough to hold it.

The commissioner, Hudson, was a hard taskmaster to whom budgets and timetables, once set, were inviolate. He pushed administrators, engineers and workers alike with punishing vigour. He was driven by a burning desire to silence the political critics who said the Scheme was too fantastic and beyond Australia's financial and technical capabilities. Working for contractors who were pushed by Hudson's ceaseless urging, tunnelling crews repeatedly broke world records for weekly progress.

Hudson was intolerant towards anyone he considered was not pulling his weight. Sackings for less than total commitment to the project were commonplace and written into standing orders to all supervising officers was the following:

> LOAFING on the job, insubordination to a supervising officer, failure to give a reasonable return of labour for a day's work, waste of the Authority's materials, or damage to plant, equipment or property, come within the description of serious or gross misconduct. In such circumstances the standing instructions are quite clear—first and second warnings before termination of employment are not required. One warning and dismissal, or in extreme cases, dismissal forthwith should be applied without hesitation.

Former members of Hudson's staff still recall with awe his anger at seeing a group of workers casually enjoying a mid-morning brew of tea by the roadside, just around the corner from head office. Such a blatant display of 'loafing' was outrageous. The commissioner said nothing but his driver, Trevor Rolling, watching through his rear vision mirror for the reaction, saw his face harden in cold fury. Arriving some minutes later at head office he marched straight into the office of his chief engineer, John 'Darby' Munro, slamming the door behind him.

About a minute later, Munro emerged, 'dark as thunder'. 'Take me down the road...you know where,' he ordered Rolling, who had waited outside, expecting such a command. The men were still lolling by the roadside when the commissioner's emissary arrived. The engineer, feet apart and hands on his hips, angrily demanded to know just what they thought they were doing. The men stared belligerently back at him: 'And who are you?' they responded. 'Munro's the name,' he said and, stiffly pointing his arm in the direction of head office, told them without further ceremony to collect their pay. He added that the commissioner himself had ordered their instant dismissal. But the men, all Australians, just stared back, bemused and clearly unconcerned.

Munro, whose sense of discipline nearly matched that of Hudson, was almost beside himself with anger. He began bawling them out as the most useless, lazy good-for-nothings it had ever been his misfortune to encounter, when one of the men climbed casually to his feet and stood in front of the railing engineer: 'Listen mate, we work for the PMG—so why don't you bugger off.' (The PMG stood for Post Master General's Department, the forerunner to Telecom and Australia Post.)

Although many thought him tyrannical, Hudson expected no more of others than he did of himself. He abhorred red tape and rejected any pomp and ceremony accorded him because of his position. On one occasion when he decided to go fishing, a well-meaning official arranged for a portable toilet to be delivered to the fishing spot. The commissioner was furious. He demonstrated his anger by standing on a rock and urinating in the open as he loudly berated the driver.

But in hindsight few, if any, would question the choice of Hudson as project leader. The government wanted a single, dynamic person in charge, rather than a board of public servants who would spend more time arguing than making decisions. He was recommended as commissioner by a senior New South Wales public servant who, at a cricket match, casually handed an envelope containing his name to the federal Minister for Works, Nelson Lemmon. Lemmon later described the subsequent interview*:

> When Hudson walked into my office he was bent on one side and wore a rather crumpled coat. I couldn't measure him up at first sight but we got talking. I said: 'I don't think you look strong enough for the job.'
> He said: 'I have a sleeping appendicitis.'
> I said: 'You won't start the Snowy job until you get it out.'

Lemmon then took Hudson to meet the Prime Minister, Ben Chifley:

> Ben sat back sucking his pipe and said: 'Well, Mr Hudson, tell me about yourself.'
> Hudson looked embarrassed and said: 'I don't like talking about myself.'
> Ben said: 'That's a pretty fair sort of an answer.' He already knew quite a lot.
> I told Hudson afterwards that the Prime Minister would back him in Cabinet.
> Hudson got his appendix out two days later.

When the appointment was discussed in Cabinet, several ministers demanded Lemmon follow Cabinet rules and submit three names for consideration. Lemmon complied and submitted Hudson, Hudson and Hudson.

After the appointment was announced, William Hudson was written up as the highest paid man in Australia. The government had decided on a salary of 5000 pounds a year, at a time when senior bureaucrats were getting half that figure and the average annual blue-collar wage was about 500 pounds.

Considering the importance of the project, it seemed a casual selection process, but

---

*From a 1964 interview with Nelson Lemmon by author Lionel Wigmore in *Struggle for the Snowy*.

Sir William Hudson, 1950.
PHOTO: THE AGE

Hudson's career had been closely studied. As well, Hudson was very keen to get the job and had discreetly forwarded his name for consideration. He was at the time engineer-in-chief of the New South Wales Metropolitan Water Sewerage and Drainage Board in Sydney and it is more than likely that the public servant who had delivered the envelope to Nelson Lemmon had received it directly from Hudson.

His wife well remembers when he returned home from the interview. He walked in grinning: 'Eileen, I've got the job, but I've had to ask Ben Chifley not to say anything until I've squared things with the Water Board.' He had taken a day off work to go to Canberra for the interview. It was the first time he had ever taken time off work and he felt guilty about the subterfuge.

Bill Hudson was born at Nelson in the South Island of New Zealand, and grew up there during the years the island's alpine water was being harnessed for hydro-electricity. He came from a family of prominent doctors and farmers and, with a Rhodes Scholar for a brother, he felt like the black sheep of the family when he broke tradition and elected to study engineering. When he revealed his intention to his father, the man, long frustrated with his son's less than brilliant school record, snapped, 'An engineer; an engineer is all you're good for.' Even when he agreed, his father still insisted on the young William going to University College, London, which was particularly known for its medical school.

In an *Australian Women's Weekly* profile at the time of his retirement, Hudson recalled his father's rigid discipline but confided, perhaps for the first time, how he quietly used to get back at the man. His father was the Nelson district's official weather recorder, and young William would often take revenge on his parent by pouring a cup of water into the rain gauge. As a result Nelson went through a period of erratic rainfall records during the first decade of the twentieth century.

Hudson went to London to study engineering and after his first year enlisted and served with the London Regiment in France. He was wounded at Bullecourt in 1917. After the war he resumed his studies, gaining a bachelor of science (engineering) degree with first-class honours and then completed a postgraduate course in hydro-electric engineering at Grenoble, France. He worked on a number of hydro-electric projects in the United Kingdom before moving to Australia for the Nepean Dam project. But the work was halted by the Depression and he was retrenched.

Not a man to remain idle, Hudson took his wife to New Zealand and left her with her family while he went to England to look for work. Ten days later, his wife received a telegram telling her to 'come to Scotland'.

In Scotland, Hudson was appointed engineer-in-charge of the Galloway Hydro-Electric Scheme, a project he saw completed in seven years, a year ahead of schedule. Out of work again—and a year earlier than necessary—the couple returned to Sydney, where Hudson joined the New South Wales Metropolitan Water Sewerage and Drainage Board.

He took up his new duties with the Snowy Scheme in August 1949, and from the outset stamped on it his energetic, no-nonsense style. Despite his rapidly earned notoriety among the workers, Hudson got the mammoth job done, like the Galloway project, on time and on budget. The last power station, Tumut-3 at Talbingo, was commissioned in 1974.

The overall Scheme comprising sixteen massive dams, eighty kilometres of aqueducts (pipelines), more than 140 kilometres of tunnels cut through the deep granite core of the Great Dividing Range, seven regional towns and seven power stations (two of them underground) cost just $820 million—about $5 billion if converted to 1989 dollars, but still a pittance when set against the estimated twenty to thirty billion dollars the Scheme would cost to build today.

Hudson was dubbed a Knight of the British Empire by Queen Elizabeth II in June 1955, the year that the first project—the Guthega Dam, tunnel and power station—was completed.

At a function marking an advanced stage of construction at the Tumut Pond Dam in 1958 the Prime Minister, Sir Robert Menzies, who by then had revised his opinion of the project, spoke of the triumph of the Scheme, and added:

> In a period in which we in Australia are still, I think, handicapped by parochialism, by a slight distrust of big ideas and of big people or of big enterprises...this Scheme is teaching us and everybody in Australia to think in a big way, to be thankful for big things, to be proud of big enterprises and...to be thankful for big men.

Many of those big men became legends in their own lifetimes. With minds confused and emotions still scarred, many needed the danger, the hard breathing from hard work and wondrous space and beauty of the Snowy Mountains.

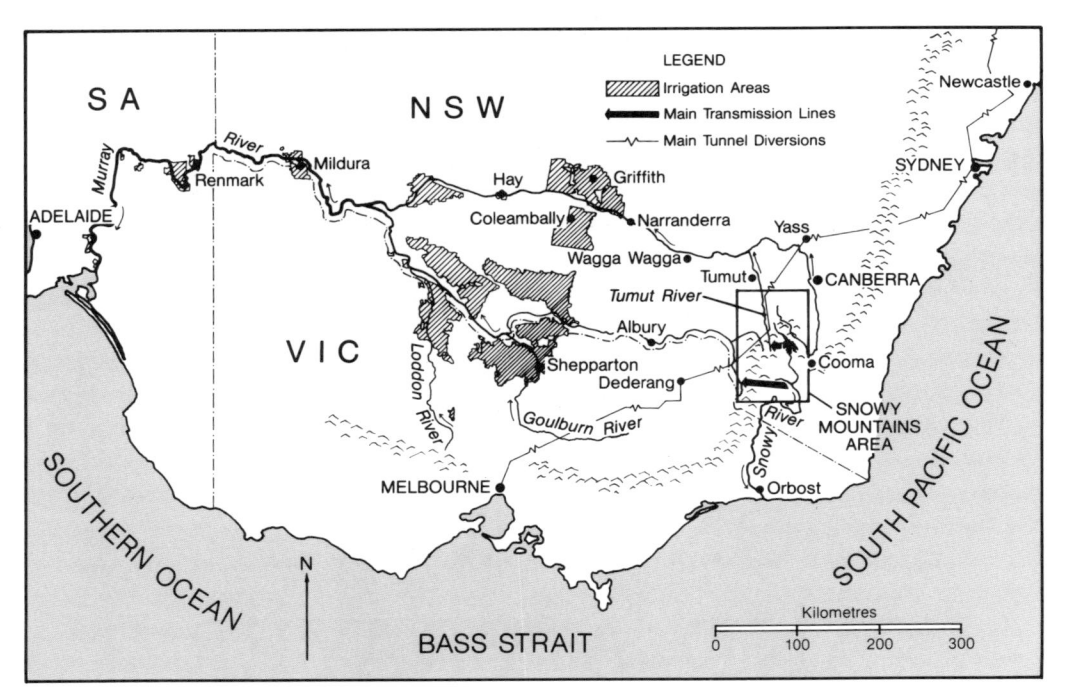

The Snowy Mountains Scheme showing the inland irrigation areas, main transmission lines and main tunnel diversions.

CHAPTER THREE

Dieter Amelung (reclining) on board the *Fairsea* on the voyage to Australia.

ON A WING AND A PRAYER

At the time federal Cabinet was gathering its thoughts on Australia's postwar future, the merits of the Snowy Mountains Scheme and immigration, a young German mechanical engineer, Hein Bergerhausen, was scouring the countryside around Cologne for expended artillery shell casings. In a back room in the house his great-grandfather had built in Frechen, a village on the outskirts of Cologne, he was cutting and shaping the massive casings into pots and pans for sale or barter. He had also constructed a beautiful, illicit brass still.

In most of Europe the postwar years were grim. There was very little food and little hope for the future. Hein had cleared the rubble in the backyard of the house into a heap; he repaired the doors and windows to make the house habitable for himself and his wife, Sybille. He then devoted his attention to the fundamentals of survival.

Hein was just nineteen when he went to war. It was soon after the German invasion of Poland and the responding declaration of war by Great Britain. Before his enlistment he was an apprentice mechanical engineer at a coalmine near Frechen. But his ambition was to specialise in aerodynamics. He had already had one interview with the Henschel Aircraft Company and had been eager to begin treading the career path he foresaw there.

All his life Hein had cherished a passion for flying. From the age of fifteen he spent every Saturday and Sunday flying a glider he and a group of friends had painstakingly built. The glider had to be dragged up a steep grassy hill near Frechen and manhandled into a heavy rubber catapult to be shot into the air with the intrepid pilot aboard. It was heady, heart-thumping stuff. Two boys were needed to stretch back the heavy rubber band, another two to push the glider into the V, while the pilot sat aboard the contraption, grimly gripping the control column and fixing his attention on the sky in front. On the command of 'los' the catapult was released and the flimsy craft hurtled into space.

All this effort produced perhaps fifteen minutes of flight. Because the operation required at least four ground crew, each boy's turn to fly came around only every two or three weeks. But those fifteen minutes of free, exhilarating flight were cherished moments. For those waiting their turn, the time was passed in raucous singing to the strains of a concertina.

They were oblivious to the political changes shaping their country. The Nazi Party under an obscure former World War I corporal, Adolf Hitler, had come to power. The only thing Hein and his friends noticed was that food became more plentiful. Hein, the youngest of eight, had vivid memories of the stories his mother and older brothers and sisters told of the 1920s when the family wandered from village to village, farm to farm, looking for food. Under the Nazis, the increased availability of potatoes, cabbages and brickettes brought a degree of comfort to lives accustomed to the austerity of the years following World War I.

They were ordinary people. They did not look for or analyse hidden political agendas. When Hitler began to rearm Germany, Hein's gliding club, like similar clubs all over the country, was told that if it registered with the government it would be supplied with materials free of charge. If the aspiring fliers thought this was good, they were doubly excited when they learned the move was designed to develop a pool of recruits for a new, modern airforce, the Luftwaffe.

When war broke out Hein was immediately drafted into this new airforce. After a six-week military training, he was transferred to Vienna for technical training as a flight mechanic. After seven months he was moved again, this time to Bavaria for pilot training. It set the pattern of constant transfers for training and increasing specialisation which, as Hein soon understood only too well, was the only thing that would keep him alive.

He had joined the Luftwaffe when few were taking the war seriously. The general opinion was that it would last perhaps six months, until Germany and England reached some agreement on disputed borders and territories set by the Treaty of Versailles. Hein, therefore, entered a formal aviation training regime that, given the rigidity of the German system, not even the intensification of the war could completely negate.

Hein was one of the lucky ones. While he went on to specialise in precision flying for reconnaissance and intelligence gathering and for low-level troop support, the eager young fliers who followed just months later were given six weeks flight instruction and sent into the skies as expendable war fodder. Hein saw boys, some of whom he knew from the gliding club, arrive after him, do thirty-six training flights, leave for a combat squadron and be killed—all before he had even gone solo.

During his advanced pilot training in Bavaria, Hein and three others, also among the original intake, were formed into a specialised unit which remained together for the whole of the war. The war was a constant cycle of training periods and postings to and fro across Europe. Every few months the unit was attached for a short time to combat squadrons—flying Henschel 129s and Junker 87 (Stuka) dive bombers—and reconnaissance units—flying Junker 88s.

The unit was led by a major who, whenever the fighting became particularly bad, seemed to have the connections to arrange the unit's transfer, either back to more specialised training, or to marginally less dangerous work such as reconnaissance. The major was a fount of information about political matters. As early as 1941 he confided to his four young subordinates that Germany would lose the war. Therefore, he told them, he would make it his priority to keep them alive.

The major was a quiet man, a philosopher and deep thinker who didn't divulge much of his past. There were even some who said he had the gift of second sight. It was not until the war's close, and the eve of the unit's dramatic escape from surrounding Russian units, that the four fliers learned that at the beginning of the war their major had been on secondment to Hitler's staff. He was, however, a flier rather than a politician. He had been careless in expressing anti-Nazi opinions and was saved only by the intervention of senior Luftwaffe officers who concocted for party officials a story of his burning desire to fight for the Third Reich. He was, thus, safely removed from the heady, insane world of the Reich Chancellery.

As a reconnaissance specialist, Hein had a window seat on some of the twentieth century's most significant events. At the beginning of May 1944 the unit was sent to Caen, France, to fly high-altitude Junker 88s, specially fitted with cameras and meteorological equipment. Its task was to gather long-range weather forecasts and record the preparations in England and along the Irish west coast for the planned second front— the invasion of Europe in Operation Overlord, or D-Day as it became popularly known.

Each day at dusk Hein took off for Norway to refuel. From there it was an exhausting and hazardous flight to Iceland, down over Ireland, across England and back to France.

ON A WING AND A PRAYER

Photographs taken by Hein Bergerhausen while flying in formation over Arad, Rumania, 1944.

A Junker 88 photographed surreptitiously by Hein Bergerhausen in 1944. Being in possession of a camera was a punishable offence. Hein had to have his film developed in France.

So that they could make the distance, the planes had been stripped of all unnecessary weight including armaments and even their exterior paint. Sometimes the route was varied slightly. In Norway the pilots were given final instructions in sealed envelopes which were not to be opened until they were airborne. The cameras were in sealed containers in the wings and were activated automatically by a timer. The procedure required precision instrument flying. It also prevented interference by anyone other than the intelligence personnel who fitted and removed the cameras before and after each flight.

Although security was important in the general scheme of things, Hein, an ordinary fellow who in any case had been told that his side was destined to lose, adopted a somewhat cavalier attitude. He had always wanted a good camera and he had no doubt that the ones he was carrying were the best. 'I always tried to pinch one, but I never could,' he later confided.

Hein was aware the navy was using U-boats to monitor the invasion build-up. The German military, too, knew the invasion was coming. They needed the photographs and weather forecasts to tell them when and where. They got the 'when' right but they were wrong about the 'where'. As the critical time approached, men like Hein were confused. Even though they were mere spear carriers, not generals, they still had eyes. It was clear to them that the army could not stop the landings but also that the massed invasion fleet would present pilots with a very large, cumbersome target. Yet on the eve of the invasion all major Luftwaffe units were recalled from the French coast. Why? The question was on a thousand lips that night.

After the war it was learned that Hitler had guessed the landing would be in Normandy, but his generals convinced him the D-Day attack was just a diversion and predicted instead the Pas-de-Calais. Despite what they told Hitler, they still withdrew the air cover from Normandy to avoid the annihilation of a large part of the Luftwaffe. Almost the entire Allied bomber command had been despatched to pulverise every German airfield within range of the beachhead.

Some of the more cynical Luftwaffe crews suggested that someone deliberately 'left the door unlocked' to facilitate the war's end. There were rumours of assassination attempts against Hitler. Even the sleep-starved mechanics failed to see how that would help. There were enough believers in the Third Reich to replace him. But with Allied troops on French soil the battlefield would steadily shrink until it withdrew into Germany itself. And then the war would finish.

Hein, however, had little time to think about this. His unit was posted immediately to Clermont-Ferrand to be trained in the detection of camouflaged Russian tanks in the Balkans. Another futile exercise to keep them occupied? Perhaps. But it would also keep them alive for a little longer, and a man still breathing had a future, even if he couldn't see it or believe in it at the time.

Before the D-Day reconnaissance flights the unit had spent weeks learning to identify Allied warships. Considering that every Allied ship afloat in the European theatre would soon be stationed off the French coast, it hardly seemed to matter what their names were. Anyone who was particularly interested could, when the time came, sit on the beach with a pair of good Zeiss binoculars and simply jot them down. Such was the

Hein Bergerhausen in May 1944.

talk of men who, though grateful to be alive, wondered sometimes if the boredom was worth it.

To the men of the Luftwaffe, the few Stukas and Messerschmitts that were left behind on D-Day created more panic than actual destruction. It was an observation that did little to improve the humour of pilots who were now suddenly on their way to the Eastern Front—a prospect that held terror for even the most hardened veteran.

As it turned out, Hein's small unit almost missed the Balkans nightmare. When they arrived at Clermont-Ferrand, the German army had moved on and the French were back in charge. What then took place was like a scene from a black comedy. The five fliers were arrested, roughed up by a few of their former allies who had decided it was time to change colours, then hauled before the town commandant. They were relieved to find the commandant was a man they knew—except he now made it plain that he didn't want to know them.

Before long the five men soon found themselves ignored and unguarded. But somewhere beyond their window a war was still raging—or was it? No one seemed to know. The men walked out into the street, asked a taxi driver which way the German army had gone and then booked him to follow it. They caught up with the army near Lyon, about a hundred kilometres to the east, paid and bade farewell to the taxi driver, and rejoined the war.

Learning that the main Luftwaffe unit they had been attached to was already in Hungary, the group, except for Hein, caught a train to Linz in Austria, and from there to Bacsalmas near the Rumanian border. Hein, however, had decided to take a few days off and go home. He had no papers but by this stage of the war, provided one kept a wary eye out for Gestapo and the military police, one could get onto a train with the aid of a hard luck story and some convincing acting.

With all the travel we had done because of our continual postings I knew how the system worked. With a bit of fast talking and some subterfuge at the checkpoints you could explain away the reason you either didn't have a pass, or why you had been delayed and gone beyond the dates stamped on the pass. It seemed to work best if you had changed into civilian clothing and explained that you had been given sudden orders to return to base, or take up a new posting, and there had been no time to receive new travel papers. That way at least you gave the impression you were returning to the war, rather than trying to desert. When you returned to base you either made up a story or, if you knew the commandant was a decent bloke, you told him the truth.

It got safer towards the end of the war when most of the guards were invalids from the front and no longer cared. But at that time, 1944, you still needed to be careful. It was up to you and how you handled every situation. You could quite easily get some bastard who would shoot you on the spot. It did happen.

Hein went home to marry his girlfriend, Sybille. It was a typical war wedding—a quiet celebration in the local town hall followed by a small reception in a friend's restaurant. The next day the couple began a brief but carefree honeymoon on a Rhine barge. They made no plans for the future; they lived only for a few days in the present.

Sybille freely acknowledged that one of the reasons she married was for her family's sake. She and her younger sister and brother had been orphaned. By marrying she would be able to keep them all together and keep her sister and brother from being placed in an orphanage: 'I was twenty-one. We were young and prepared to just hope for the best.'

During the second half of 1944 and in early 1945 Hein's unit was used to patrol and keep open the main railway links for troop movement. The group's reconnaissance expertise was also used to seek out hidden Russian tank divisions, searching for the telltale darker shadows in the forests. They photographed all farmhouses and outbuildings big enough to conceal a tank and the photographs were scrutinised for scuffed ground around the buildings.

In February 1945 the unit was recalled from Bacsalmas to Linz. Hein's five-man group, however, was detached and sent to Lübeck in northern Germany to train on the first Messerschmitt-262 jet fighters. But there was confusion and, for the first time, signs of panic in the lines of command. The wheels were falling off the great war machine. In between learning to handle the latest technology, the fliers were set to work digging trenches and foxholes in the hard, black earth.

Six weeks later, before the men had even grown used to the new smell of burning kerosene, they were sent back to Linz. Why? No one asked. No one cared any more. In Linz they found that half their comrades from Bacsalmas had been dispatched to a suicide commando, flying Messerschmitt 109s packed with explosives. Their orders had been to fly into the midst of bomber flights, set a short fuse and bale out. It was a last, futile action by commanders who still believed in the politics of what they were doing.

One wonders if it would have made any difference had the men known that Adolf Hitler was now clearly insane and had been living for months in a concrete bunker in the garden of his Chancellery; a broken man with trembling hands, twitching arms and a dragging leg; a 'stooped figure with a pale and puffy face', who stared through glazed eyes as his dream of a 'New Order for Europe' crumbled about his head. But the waste of life continued.

Hein and his friends were deeply shocked and for the first time felt an all-consuming fear. They knew of no one who had survived these insane missions. With the help of their major, they were transferred out of this nightmare to Berlin, where they joined the command of Oberst (Colonel) Hans Ulrich Rudel, one of the war's most highly decorated German fliers. Under Rudel they were given the task of setting up an anti-tank unit to be deployed against the Russian advance, both in eastern Germany and, later, in the Balkans.

With the Russians almost at the outskirts of Berlin, Hein's unit was moved to Czechoslovakia to try to help stem the advance there. But it was too late, and while the generals sued for peace and prepared to write their memoirs, the rank and file were left to find their own way home. For the foot soldiers on the Russian front, it was a hopeless undertaking. For tens of thousands of hungry, homeless, defeated soldiers, the end of the war meant the beginning of slavery. They were not returning home to the mothers, wives and children who, after surviving the bombing, waited hopefully in the rubble of their past.

The night before Russian troops were expected to close on the airfield, the unit made hurried plans for its escape. But the few remaining Stukas and Messerschmitt 109s were not enough to evacuate all the ground crew. The Stukas were stripped of all unnecessary equipment, including the wing-mounted machine guns. By stripping the rear cockpits each plane was made capable of carrying three people instead of the usual two. But even then there was not room for everyone. Some men had to be left behind and it was decided they would try to make their way on foot through the Russian lines. The airfield was surrounded by forest and the German border was only fifteen kilometres away. If they left that night they stood a fair chance of making it back into Germany.

The major, the philosopher and father figure who had protected his young charges during four years of mayhem, volunteered to stay with the ground crew and help organise their escape. Frightened and sad, the men farewelled each other. When Hein said goodbye to the major they simply shook hands; there was no time for emotional goodbyes. But the major offered Hein advice that was to guide him for many years: 'It will be hard when you get back home. It is important to forget what it was like before. Look after your family and seek work—any work, even if you have to do it for nothing. Hard work is what you will build your future on. Farewell!' The moment remained etched like a photograph into Hein's memory. He never saw the major again. When he inquired after the war, the man had vanished without trace.

The pilots and their passengers waited until the next day to leave, hoping to get final word of the status of the war. They went on a short observation flight to check on the Russian advance and watched helplessly as the Russian army steadily overran the fleeing German troops. The patrol returned to base, now the centre of a shrinking island in a Russian sea. It was late in the morning of 8 May 1945. Waiting for the pilots was a telegram from Berlin telling them that all action must cease at twelve noon. Hein prepared for his return home by stuffing warm clothing into the cavities where the machine guns had been pulled out. He did not know what awaited him, but he expected that warm clothing would be at a premium. At about twelve-thirty the men shook hands and went to their aircraft. Hein, who had the longest journey home, was the first to take off.

As he lifted the lumbering Stuka over the treetops he passed over Russian troops and armaments, the red painted stars flashing beneath like the upturned eyes of alien insects. The Russian weaponry drifted under the fuselage in terrifying slow motion but no shots were fired. Hein prayed for the time when the stars on the vehicle would turn to white, signifying that he had crossed into American-held territory. He was frightened of being caught by the Russians and pondered the absurd irony of being shot on the last day of the war.

If all went well and he could find somewhere safe to land and refuel, Hein intended to take the Stuka all the way home and put it down on the old grass strip that he and his boyhood friends had used for their glider. There he would abandon the plane, and the war. He would simply walk home with his passengers, who also came from near Cologne.

As soon as he had left the Russians behind, Hein descended again to treetop level, hugging the contours of the land and following a river. He hoped that by keeping in the shadow of the valley he would be out of sight of patrolling Allied aircraft. The theory was fine but the end came from an unexpected direction just two hours into the journey. A hilltop anti-aircraft battery crew had seen him coming through the valley below and had been able to traverse their gun low enough to train the sights into the valley. When the ack ack exploded around him, Hein was bewildered. He was probably the first German pilot in the war to be shot at from above by anti-aircraft guns. His passengers, who could see what was happening, undid their harness and reached forward to pummel the back of his neck with their fists and the emergency canopy winder. They yelled at him to land; but there was nowhere to go. The hill and its gunners were straight ahead. Hein marvelled at how pretty the muzzle flashes looked as he flew into them.

Trying to ignore the beating he was receiving from his distraught passengers, Hein lifted the Stuka over the hill and battery, and flew straight into the waiting 'arms' of a pair of American Thunderbolt fighters. He flung back his canopy and fired red, green and red flares—the agreed surrender signal—in quick succession. 'We waggled wings at each other, and the war was over. They were bloody careless though. They flew in front when they should have stayed behind me. If I had had cannons I could have shot them both down—not that I would have at that stage of things.'

Hein followed the Americans to a base they had taken over, and on landing he stared with amazement at the first black soldier he had ever seen. The man seemed to appear on his wing from nowhere the moment the aircraft trundled to a stop. The American leaned into the cockpit and souvenired Hein's wristwatch, his Zeiss sunglasses, his silver pilot's wings, a silver bracelet given to him by Sybille when they married, and his service pistol.

But he barely noticed being frisked. He just felt a huge weight lift from his shoulders. 'I vowed there and then that I would never fly again. I couldn't believe I was still alive. So many friends were dead.'

The three men in the Stuka were taken for a brief interrogation, which sought only formal information about identity and where they had flown from. That night they were billeted in unguarded barracks and invited to drink with the Americans. Looking at their faces, Hein thought how young they all looked. Few, it seemed, had seen combat. Hein, at twenty-five, had the lined face of a veteran. The Germans were asked if they had any complaints about their treatment. Hein mentioned the loss of his belongings to the black soldier. Within five minutes everything except his silver wings and pistol was returned.

That night a German officer called his countrymen together and advised them to tell the Americans all they knew. The war was over, he said. They should now accept the authority of the American officers and obey their orders.

The next day the tempo changed dramatically. The interrogations were handled by senior officers and lasted all day. Hein's group had the most up-to-date information on the Russian positions and the Americans were very interested: 'How many tanks did you see? How are they dug in? Show us the location of German fuel and weapons dumps...'

The following morning Hein and other captured pilots and ground crew were ordered back to their aircraft to make them ready for flying. The aircraft were rearmed and with cold dread the men realised they were being readied again for war. They learned from the Americans that orders were expected at any moment to push on against the Russians.

It seemed the world really had gone mad. The Germans were now being conscripted by the Americans who, it seemed, had yet to satisfy their lust for battle.

But by nightfall the generals and politicians had apparently decided otherwise. For the Germans it meant an early morning journey, by road, to a prisoner-of-war camp—an open air compound where interrogations resumed in earnest as the Allies began a meticulous screening to seek out Nazis and war criminals. There were thousands to be processed and the only food provided was a piece of cheese and a biscuit twice a day. 'But we didn't complain. No one complained because we knew they just didn't have anything else; though a few of us pinched some old bits of furniture and a few potatoes which we were then able to roast on a small fire.'

To get the cheese and biscuits the prisoners had to file from one enclosure to another. As they did so their faces were studied closely. Twice a day, there and back; four times a day for several days every man was scrutinised. People who could identify wanted men were helping the Allies in their scrutiny. Hein was in the camp for three days before he was individually interrogated, his papers cleared and he was given permission to return home.

Twelve days later he was in the rubble that was Cologne. He used a three-day break provided by a weekend to clear rubble and make his old house in Frechen habitable for him and Sybille. Sixteen days after the war had ended he was back working in the mine as a mechanical engineer. His boyhood dreams of a flying career had turned sour.

In his spare time Hein used his skills to help others rebuild. He set to work repairing people's pots, pans, door locks; and cutting and hammering his clandestine collection of artillery casings into new utensils. In return, people would offer butter or bread or vegetables. The community was healing itself and starting afresh.

But there were some scars which ran so deep that people wondered if they would ever heal. As the population drifted back and resettled there were houses in almost every street that were never reclaimed. Their owners, old neighbours and friends, had been swallowed up, gone without trace. It was only when the Nazi war trials began that the truth was revealed in all its horror. People began to remember: 'Yes, they were Jewish... or communists...wasn't he a trade union leader?' These had been ordinary German families who, because of their beliefs, were deemed to have no place in the 'New Order for Europe'. They had been exterminated. The Holocaust was not something people spoke about, but it glared out at them every time they walked past an empty house.

People, however, are easily distracted by the concerns of daily life. The centre of Hein's attempts to rebuild his life was the big brass still he constructed from artillery casings. The black market schnapps he made was his only real source of cash. At the mine the workers were paid in goat soup, brickettes and enough cash to buy seven Camel cigarettes a week. Fuel brickettes and cigarettes could subsequently be bartered for almost anything. 'But with the still we soon ceased to talk of single Camels; we thought in terms of cartons.'

A cook at a nearby American army base, whom Hein befriended, surreptitiously supplied him with drums of fruit salad. The fruit salad made exquisite schnapps: 'Tropical and Californian fruits I had never seen before. Beautiful! I had so many friends, you wouldn't believe it. But most of it I sold to the US army. Sometimes we'd run a bit short—no fault of the management—so we'd add a bit of tea to top up the bottles. As soon as we were around the corner after making a sale we'd run like hell.'

Business was good, but the little backroom distillery was a demanding beast. Nights were the only time Hein could tend the still, and in order to get some sleep he made a float level which activated an alarm clock when the drip flask was full. In the morning he would restart the process and leave Sybille in charge of the glistening gurgling contraption.

As time passed, men began to return home from the prisoner-of-war and labour camps set up by countries which had laid claim to German soldiers for war reparation. Hein's illicit brewery became the focal point for the village: 'We partied every night. We had nothing to eat and nothing to wear but we were young and the war was over.' For two years the still did a brisk business that helped Hein and Sybille back on their feet. But they always knew it was too good to last.

One day Sybille's sister, Sophie, telephoned Hein at the mine: 'Come home, quick. The customs men have been here. They are looking for a still.' Sophie knew all about the still. She had helped tend it often enough but had feigned ignorance. Besides, she told the customs inspectors, she didn't have a key to the back room, where they were so convinced it was. If they would like to return later, she said, she would go to the mine and get the key from Hein.

Hein sprinted home with a little under an hour to spare before the inspectors were due back. He dismantled the apparatus and hid it beneath the rubble in the backyard. When the inspectors returned and Sophie let them into the room it reeked of alcohol. But the still was gone, forever.

By 1949 when Major Clews was organising his survey teams in the Australian Alps, Hein Bergerhausen was coming to the conclusion there was no future in Germany. He began to reflect on his years of flying to and fro across Europe, not in terms of war and stress but rather with a sense of freedom lost. Peace was hemming him in. The house was overcrowded with both his and Sybille's families. He remembered his major's words and knew it was time to leave.

He had started learning Spanish and was sitting in a waiting room before his third and final interview for work in Venezuela when a priest asked him his business. 'Think again about Venezuela,' the priest warned. He disclosed that German tradesmen were being recruited for its oilfields, in malaria-infested swamp country. 'They will offer you a new start, including a new name. But if you then try to leave they will use the change of name as a lever and threaten to allege you are a war criminal.' Hein went into his final interview and was offered a new name. He declined the job.

A year passed. Then, early in 1951, a friend, a local police detective, invited him to ride pillion on his new motor bike. He was going to try it out on the motorway to Düsseldorf, where he had succeeded in getting an interview for work in Australia. 'They are building a big hydro-electricity scheme there and are looking for workers,' he said. Hein was barely listening. He had to cast his mind back to school geography to remember where Australia was. But he was keen to ride the bike because the motorway, built during the war, would have little traffic. They could fly its smooth, deserted lanes.

The policeman had the last interview of the day, leaving Hein alone in the waiting room. Hein was gazing aimlessly at a blank wall when a man approached. He asked if he was there for an interview.

'No, I'm just waiting for a friend who is,' Hein replied.

The man held out his hand and introduced himself. 'Walter Hartwig,' he said. 'I am an engineer. I am helping with the recruitment for the Snowy Mountains Authority in Australia. We are going to be building many dams and power stations. Perhaps you have heard about it? We had a broadcast on the radio.'

Hein shook his head.

'Well, what do you do? What is your occupation?'

'Mechanical engineer.'

Hartwig said skilled people were in strong demand: 'The Australian government gave first option to workers from our lost eastern provinces—Prussia, Silesia—but it's not enough. We're looking for good workers and tradesmen. Men who want a challenge, a new start; men who are content to work and settle down. Sound like you?'

Hein shrugged but was becoming interested.

'Have you thought about going to Australia?'

Hein shrugged again. Such a thought had never entered his head. He was only up for the motor bike ride.

'Come and have a chat then.'

Before he knew it, Hein was being ushered into an interview room where another engineer joined Walter Hartwig. The two men probed Hein's background and qualifications. They were testing, looking for his willingness to learn another language, English. They studied his appearance, his willingness to cooperate, and his frankness.

All applicants were being carefully screened. The Australian government, at that stage, would not accept anyone with Nazi links or sympathies. The second engineer splayed his hands on the desk. 'A lot of good tradesmen have had to be eliminated because of their political background—Hitler Youth and so forth,' he said, a touch of resignation in his voice.

They then told Hein about the Scheme. He was surprised to learn there were other recruiting centres in West Berlin, Hanover, Hamburg and Bremen. He began to realise how big this project was. Because of the widespread unemployment and the economic depression that was racking the country, the German government had refused permission for the Snowy Mountains Authority to advertise in newspapers for fear of general unrest around recruiting centres. But the Authority's agents in Germany noted that the order mentioned nothing about radio. They made a single broadcast on a current affairs program and within a week had been swamped with more than 10 000 applications.

The two engineers told Hein that if he was accepted he would be going where there was no civilisation: 'It is in the mountains. It will be lonely. You will be starting from the beginning. Could you cope with such conditions?' Hein didn't reply; he was thinking very hard. A door had opened revealing prospects that excited him but which he could not yet describe.

Unknown to Hein, the man interviewing him, Walter Hartwig, had had the same feeling. He too was eagerly looking forward to a new life in Australia. The Nazi experience had left a bitter taste in his mouth. He would never forget the intimidation and fear generated by the Brown Shirts, or his wasted years as a prisoner of war.

Walter had been a sergeant, a heavy machine gunner in an infantry unit. In April 1943 he was near Tunis. One day, after a major battle, he and a corporal became separated from their unit. They did not know where they were or how to reach their own lines. In the distance they caught sight of a column of scout cars with black pennants. 'Good,' Walter had said to the corporal. 'Italians. We can get a lift with them.' He began to wave, then realised they were British.

> We quickly ran and hid in a cactus field but had no sooner found a hiding place than the air about our heads was filled with the blast of heavy machine-gun bullets. I looked around and found myself staring into the muzzle of a machine gun on a scout car which had seen us and circled behind.
>
> I waved my hat in surrender. 'Come on,' I said to the corporal.
>
> 'Oh no, sergeant, you first please.'
>
> He wasn't being polite. We didn't know if they would shoot us or not. Holding my hat high, I walked towards the scout car and we were taken prisoner by a sergeant-major. I often wished I could have met the man after the war and thanked him. I'm sure he had stopped the gunner from killing us.

The pair were subsequently handed over to the Americans and Walter spent the rest of the war as a prisoner of war in Arkansas.

The political zealotry which had been responsible for the war had made a strong impression on Walter Hartwig. It terrified him. As he interviewed prospective workers for Australia he probed and goaded, looking for the slightest signs of fanaticism. He understood enough about Australia to know it was a young country populated by easy-going people. Prompted by his own experiences, he set out to ensure that no worker from Germany arrived there carrying with him the seeds of mindless nationalism and prejudice.

Never again did he want to sit at the dinner table with his family, waiting in fear for the knock on the door as Brown Shirts came for their party contributions. They spared or cared for no one. If a person did not have the money, he knew what to expect: to lie awake at night in fear, waiting for the stones to smash through the windows; or suddenly and mysteriously to lose his job; or even worse. One did not have to be Jewish to have one's teeth or arms broken—or simply to vanish.

Walter Hartwig had an upper middle class background, a social group that despised the Nazis. He was educated in a college system the Nazis banned, and so was acutely aware of how easily prejudice and fanaticism could flare and spread until they engulfed an entire society.

At the end of the war he returned to Germany and gained a position as regional manager for a large construction company. But he was uncomfortable living in a country that had destroyed itself and most of Europe. Although he had been raised in the traditions of a patriotic German family, he understood the vast difference between patriotism and nationalism. After reading an article in an engineering magazine, Hartwig wrote to the Snowy Mountains Hydro-Electric Authority. When he was offered the chance to join the Scheme and live in Australia, he grasped it with both hands.

A week after Hein Bergerhausen's unplanned interview with Walter Hartwig, the former Luftwaffe pilot received a letter formally offering him a two-year contract with the Snowy Mountains Authority. The fare of 156 pounds would be deducted from his wages. Wages! They were three, four times as much as he could expect to earn in Germany. He was asked to produce a certificate from the German Police Department verifying he had no criminal record; and to have x-rays taken to satisfy Australian health authorities that he had no tuberculosis.

Hein was more excited than he had been for many years. Sybille was overjoyed, even though Hein would have to go alone, at least in the beginning. But it represented a new life. It was like something from a dream.

They had been worried for some time about the deteriorating political situation in Europe. It seemed that any day the United States and Soviet Union would resume the war and use Europe as the battlefield. 'I used to lie awake at night and wish that just one American city would be bombed like Cologne, so there could be peace forever. The American people had not felt war, and that frightened me more than anything,' said Sybille.

The news was not good, however, for Hein's friend the policeman. His letter from the Authority explained, apologetically, that Australia had enough policemen.

At the end of July 1951, Hein boarded MS *Skaubryn* in Bremen harbour. There were sixty men to a cabin, stacked in three-tier bunks. They were all bound for Melbourne, and thence to a place called East Camp in a town called Cooma. No one had been able to find Cooma on a map, but it was a happy ship that sailed from Germany's shores on the last day of July—even though sailing had been delayed for two days because of a strike. Once on board, the German migrants were told they would have to work as part of the crew. This was the arrangement for the displaced persons from the refugee camps, but the Germans argued they were being charged full fare. After talks between the Snowy Mountains Hydro-Electric Authority and the ship owners it was agreed that on-board work for the Germans would be voluntary. The principle established, the Germans turned their hands enthusiastically to seamanship.

Everyone, it seemed, was on the move; in shoes that flapped, on bicycles that squeaked, and in old wooden carts that jolted and rattled. Bedraggled but treading with purpose, an entire population of ragamuffins seemed to be heading home—or looking for a home.

The skinny urchin, his scrawny neck and shoulders tugging painfully towards the dusty, pitted road under the weight of the two suitcases he grasped in his bony hands, walked almost unseeingly. Dust, congealed by sweat that had trickled around the thin wire frames of his spectacles and blurred the lenses, isolated his senses from the throng.

Malnourished, and tortured by the pain in his arms and feet, he had abandoned all rational thoughts, except for the need to keep heading east. There, almost 400 kilometres across the roof of Italy and beyond the Gulf of Venice, was home—a beautiful green valley of villages and vineyards in a region called Slovenia in the north of what is now Yugoslavia. It was a long way to walk with no food and little money, but it was better than the slow starvation he had left behind.

Ivan Kobal had not seen the little hillside village of Planina since the start of the

The *MS Skaubryn*, used by the Snowy Mountains Authority to bring German tradesmen to Australia in 1951.

war when, as one of a chosen few, he had been sent to Italy for higher education. It was one of the Fascist regime's small concessions to a people rebellious over the imposition of a new language, Italian, and the suppression of their own Slovenian culture. Ivan had accepted the education, nurturing a vague adolescent idea that he might one day be able to use it against the oppressors.

From behind the walls of the Jesuit college he attended at Piacenza, in the north-west of Italy, the war seemed almost unreal. Battles and the fortunes of armies became coloured pins on classroom maps, their movements dictated by propaganda bulletins. They were abstract events, far distant from the disciplined routines of study and prayer. War touched the boys only in snatches. At Christmas 1942, the students shared their teachers' whispered anxiety about the safety of Pope Pius XII after his public condemnation of Fascism and Nazism. In 1943 they wondered at the arrival of German troops to take control of district administration. As winter ushered in 1944, their stomachs tightened as increased austerity affected mealtimes.

One day they were given a more direct glimpse of the reality of the war. They were assembled in the bitter cold courtyard and surrounded by armed soldiers. They watched and listened in trepidation as the college principal argued heatedly with a German officer who was demanding to search for partisans he insisted were being sheltered. The two men's voices cut the still, frozen air, their faces almost obscured by the vapour of their breath. The boys felt the tension scratch their spines. For all they knew, there could well have been partisans hiding in the grounds. But, to everyone's surprise and relief, the principal won the argument and the troops filed back through the large iron gates, back to their war in the outside world.

Later, the boys heard the first bombers drone through the night sky. They sat on their wire-sprung bunks, eyes peering into the dim, high ceiling, wondering what the noise meant. But these were just distractions from what was otherwise an enforced preoccupation with education. The war did not bite into their lives until its final months, when the chaos outside was too overwhelming to be kept out by stone walls. There was no food and some students were growing so weak they needed medical treatment, which could not be obtained. There was no civil order. Normal routines were rendered pointless. Teachers began to leave. Some were taken away; no one was sure by whom or where. Students, too, began simply to leave. Eventually all older boys were asked to go, 'to take their chances'. Ivan, now seventeen, was one of those. He packed his belongings, walked out the gate, and kept walking.

On his third night on the road he staggered into a railway station after overhearing that there might be a train to Venice later that night. He found space against a wall on the platform, lay his suitcases flat and curled up, exhausted, on top of them. When he awoke some hours later, there was nothing beneath him but the cold, dirty cement of the platform. Everything he possessed in the world was gone, including a prized Columbus fountain pen which he had won for Latin and which he had wanted to show proudly to his mother.

The loss hurt deeply, though it did at least mean he was no longer burdened by his suitcases for the remainder of the journey. The following days passed in a blur as he trod through ruined towns and cities until he was walking the final dusty road into his valley. But while he was away at school, the border had changed. Just one kilometre, within view of his home, he was stopped by border guards. His papers had been lost with his suitcases. It took more than an hour of explanation and pleading before Ivan was finally allowed to pass. 'I had come home to see how life had changed now we were free from Fascism and this man was stopping me entering my own valley. I felt then that the changes might not be what I had hoped for.'

He joined the family and the people of the valley in their attempts to recover and rebuild. Everybody was scarred. Ivan had arrived home suffering from severe nervous exhaustion. His father had arrived only shortly before him, a walking skeleton, after three years in a slave labour camp in Germany. His father's best friend, a neighbour in the valley, had been put into a furnace, just a week before liberation, when he became too weak to work.

On the day he arrived home, Ivan's father was approaching the family cottage when he saw his elder son in a field, showing two German prisoners how to use a scythe for cutting hay. The young man had put aside the rifle the authorities had given him to guard the prisoners. The sight of the abandoned gun and the trust his son was bestowing on the German soldiers so enraged the man that he snatched the weapon and trained it, not on the Germans, but on his bewildered son. The Germans, simple soldiers from Berlin, stood silent and anxious as the man unleashed a torrent of anger and hatred at them, and abuse at his son.

The valley had been freed from Fascism but the new peace was not what many had longed for. As the family regrouped a new wave of political fervour began to sweep the land. The Communist Party was using force to weld Yugoslavia's ethnic factions together. Party meetings, which all were expected to attend, were held across the country as the new regime outlined the way people were to live. It was not long before the new rulers stamped their authority on the valley. Ivan, as a young man of education, was marked and watched.

For a while he was too preoccupied with recovering his health to be concerned with politics, but a year after the war the people of the valley were ordered to gather to honour a partisan hero, killed fighting the Germans in 1942. The meeting was to be held at a shrine to Our Lady in a remote natural garden. Sixty thousand people attended and when they took up a chant glorifying Stalin and the new Yugoslav leader, Tito, Ivan felt his stomach turn. The crowd became hysterical with the catchcries they were fed and a deep foreboding gripped the young man.

> I was overwhelmed by the mindless mass hysteria. I knew the people were shouting against their own private convictions. They were doing it because they were frightened. Having won the war against the Fascists I thought we would now be free people. Instead we were being told how, this way and that, we had to continue the revolution. It became necessary to be very careful how we behaved and what we said. This wasn't the freedom I had imagined and I couldn't stay neutral. There weren't many people who had finished high school and I knew the authorities were planning to use me to their advantage. But I couldn't agree with their doctrine. My studies had marked me and I knew I had to leave—immediately.

That same night, Ivan told his family of his decision, saying that, as he had little of material value to restrain him and as he had made up his mind, he would leave straight away. Though the border was only a kilometre from the farm, frequent patrols made crossing dangerous. His brother contacted a friend who knew the nearby forest well and would take Ivan across the border for a small fee.

It was January 1947. Winter had set in. The night was bitter but dark and the pair crossed into Italy without incident. Ivan had 130 lire. He gave half to his guide and spent the rest on a bus fare to Gorizia, the nearest major town, where he hoped to stay with a Jesuit community. But his contact, a local priest, had died. His replacement urged him to move on quickly, either back across the border or deeper into Italy.

Yugoslav agents were known to be active in the region, looking for people who had crossed over. Those caught simply disappeared and it did not require much ingenuity to imagine their fate. So Ivan returned to the road, finally reaching the city of Padova, west of Venice and about 140 kilometres from the border. There he found work as a kitchen hand, but soon learned through friends that even Padova was too close to the border. Too many strangers were asking too many questions, and too many people were vanishing in the night.

Ivan was faced with two options: to try to live completely as an Italian, or submit himself as a refugee. Deciding he could not pass as an Italian, he used the few liras he had earned to catch a bus to a refugee camp that had been established in a former army barracks near Naples. He remained there for three years, a period he described as the darkest of his life.

Like so many European refugees in camps scattered around the Continent, Ivan had started learning Spanish in the hope of eventually making his way to Argentina. The arrival in the camp of an Australian delegation looking for 'good pick and shovel men, young and strong' was unexpected. He wasted no time applying and switching his language course from Spanish to English.

When he was accepted by the Australians the darkness lifted, the malnutrition no longer seemed quite so severe and the weeks suddenly passed in a rapid, blinding blur. He retained only vague memories of the squeeze with hundreds of others into makeshift cabins on the *General Sturgess*, a decommissioned American warship; of breathing lungfuls of fresh sea air after a night among hundreds of sweaty bodies; and of the overpowering excitement of standing among the lean men of a dozen nationalities on a wharf on the other side of the world, in a city called Melbourne.

In 1948 Dieter Amelung was trapped in a divided nation. He was in the Russian sector of Germany; the rest of his family was more happily stranded in northern Bavaria, on the western side. The Wall had yet to be built, but crossing the heavily patrolled border was still fraught with danger and was a move to be considered only as a last, desperate resort.

When the war ended and the Allies and Russians carved Germany into two, Dieter tried to live with the change. He left school and studied to be an electrician. But with his training behind him he began to think more deeply about his situation. Life seemed

Survey camp at Reid's Flat near Indi, 1953.

The Geehi Dam camp, 1958. Geehi Dam impounds water diverted by the Snowy-Geehi tunnel and from the Geehi River, providing the headwater for Murray-1 power station.

The mud brick cottage Major Clews built for his retirement on the site of his old camp at Indi. It was re-conditioned in 1988 by members of the army survey corps as a bicentenary project.

Opposite

The site of Island Bend township today.

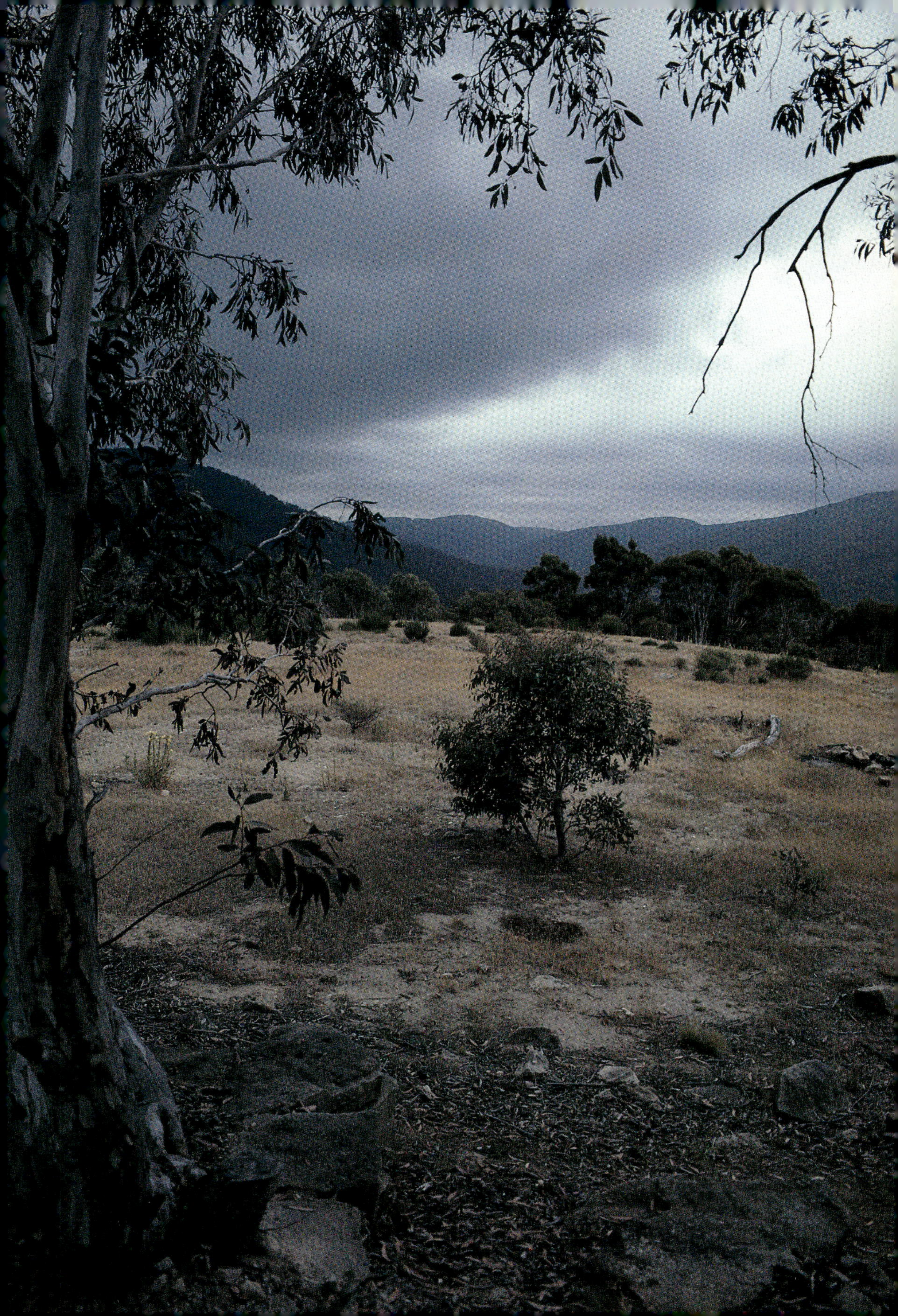

Island Bend dam impounds flow from the headwaters of the Snowy River for diversion through either the Snowy–Geehi tunnel to the Murray power stations, or the Eucumbene–Snowy tunnel to Lake Eucumbene and the Tumut scheme.

to be getting steadily worse under the Russian occupation. Then one day the Americans restructured and revalued the West German currency, plunging East Germany overnight into an economic wilderness. This dug the foundation trench into which the Russians built the Wall.

The hostility between the two superpowers seeded dread in the minds of a people still coping with the aftermath of war, and Dieter joined the tens of thousands who became obsessed with the idea of escaping from the enclosing oppression. Finally, he decided to 'vote with his feet' and caught a train to the north where, he had heard, it was easier to cross over. He alighted at a small one-platform station west of Halberstadt, where he saw dozens of people leaving the train and making their way into the surrounding forest to hide until nightfall. Dieter watched them in disbelief. They were so obvious. There were certain to be patrols through the forest.

Feeling that subterfuge would be a waste of time, Dieter asked an old man on the platform where the border lay. The man looked at him grimly and pointed north-west along the railway tracks. Dieter thanked him, climbed down onto the tracks and followed the thin ribbons of steel. He had been walking for about thirty minutes, feeling more relaxed with each step and wondering how he would recognise the border, when he rounded a bend in the line and came face to face with two armed policemen.

To run would have been futile. He kept walking, lifting his suddenly leaden legs from sleeper to sleeper. He called a greeting to the policemen, who stared at him stolidly as he approached. He spoke again as he reached the men and when they responded he recognised their dialect as that of his own area, near Leipzig in the south. It gave him confidence. Avoiding any reference to the question of why he was there, he talked about the familiar places and better times that he and the guards had known. They offered him a cigarette; no one wanted to face the situation and the decisions that now confronted the three of them.

When the last of the cigarette smoke was exhaled, the conversation stalled. One of the guards gazed at Dieter and asked him his name and how old he was. 'Dieter...eighteen.' The guard pointed to the forest on his left. 'Well Dieter, we're going *this* way.' He pointed along the railway line in the direction Dieter had been heading. 'If you keep going *that* way, you will find the border.' Before Dieter could fully absorb what the man was saying, the guards walked away.

Just a few hundred metres further down the track, Dieter crossed a small stream and saw another guard, who studiously ignored him. But it no longer mattered. The uniform was that of a West German border guard and Dieter knew he had crossed.

He left the tracks and cut across country towards a small village he could see in the distance. He went straight to the railway station and caught an overnight train to Hanover. There he boarded an express destined for Frankfurt. The carriage was jammed with people and he was forced to stand, but he felt exhilarated. He was in the West and on his way to rejoin his family, whom he had feared he would never see again.

About two hours into the journey the train began to slow. Peering through a piece of window, Dieter could see nothing but forest. A feeling of foreboding gripped him. With a cacophony of screeching metal and hissing steam the train stopped. Inside the carriage people were looking about with alarm. Peering again through the window, Dieter saw with disbelief the station from which he had begun his walk to freedom the previous day. This time, though, there seemed to be Russian troops everywhere.

The country's roads and railways had yet to accommodate the new border. About seventeen kilometres of the Hanover–Frankfurt line curved through what was now East Germany, giving the East German authorities a second chance to ensnare hapless fugitives.

Two policemen squeezed their way into the crowded carriage, calling for passports and identity cards. Dieter's identity card was East German. As he watched the policemen working their way down the carriage he knew he was facing long, perhaps indefinite, years in a prison. Terror gripped him. He tasted its acidity; felt it burn and dry his quivering mouth. Suddenly one of the policemen was next to where Dieter stood. In a brazen effort to keep his tenuous hold on freedom, Dieter snapped out of his mental torpor. Caught in the crush of passengers, he feigned annoyance at being forced yet again to extract his papers. He brusquely told the policeman he had already shown his passport to the other officer. With a contemptuous glare, the policeman thrust forward his hand. At that moment an older man beside Dieter came to his rescue. 'That is so,' he told the policeman. Just then, an even greater crush, caused by the movement of the other policeman, forced the first one to struggle on. The second policeman also asked for Dieter's papers, but this time it was easier to claim he had already shown them. The policeman moved on.

To Dieter, time seemed to freeze, refusing to move until he was a prisoner. He was sure the two policemen would return. When they came, the sound of escaping steam and the jolt of the carriage were the sweetest sensations of his life. The man beside him put a hand on his shoulder and whispered quietly in his ear, 'You are a lucky boy, eh?'

Two days later Dieter was with his parents in the small northern Bavarian village in which they had settled after his father's release as an American prisoner of war. There were many refugees from throughout Eastern Europe and the villagers made it plain that they were not welcome. Although he found a job locally as an electrician, Dieter felt like an outcast: 'I was a refugee, a stranger, in my own country.'

He steeled himself to be patient, but a family friend who had emigrated to the United States before the war arrived unexpectedly one day and regaled Dieter with stories of the good life in America. He harboured these thoughts for a year, until the increasing tensions between America and Russia made him believe another war was inevitable and that he should leave the country. He decided to try to emigrate. He saw a poster in the local employment office promoting Australia. He applied to go there, but was rejected because 'they had no demand for electricians'.

Dieter next applied to migrate to Canada, and was accepted. Some weeks later, on the day he collected his Canadian visa, a letter arrived from the Australian authorities asking him if he could return for another interview. He did, and was accepted. Two months later he was aboard a ship bound for Melbourne.

On board, some of the Germans were talking about work on a hydro-electric scheme in the mountains. As an electrician Dieter was interested but, as no one had mentioned it to him officially, he thought no more about it.

Dieter Amelung farewelling his parents in Germany at the start of his quest for a new life on the other side of the world.

CHAPTER FOUR

Sharp Street, Cooma in 1950.

JINGLE OF BIT, CREAK OF LEATHER

By the late 1940s, the outside world had made only superficial impressions on the lives of the people and 'the way of things' on the Monaro Plains below the Snowy Mountains. Home of the high country graziers who gave birth to legends like the 'The Man from Snowy River', it was one of the last bastions of traditional 'outback' living in southern Australia. The people of the Monaro held stoically and fearlessly to the values of an era that typified for many the source of the true Australian psyche.

Cars were few, sealed roads even rarer. The occasional passage of modern transport across the vast expanse of open grazing country could be tracked from afar by the plume of opaque brown dust which hid, even at close quarters, the vehicle responsible for it.

The Monaro, discovered in the 1820s by explorers in search of grazing country to sustain the fledgling but rapidly expanding wool industry, was one of the first areas settled by squatters. It was a natural plain, its deep basalt rock forming an impenetrable barrier to most native trees. It was this feature which gave it its name. Phonetically it had been the 'Manaro', or 'Maneroo', ever since, tens of thousands of years earlier, it became known to the Aborigines. When the squatters learned its Aboriginal name, meaning 'treeless plains', they saw no need to change it.

Against the deep violet backdrop of spectral mountains it lay as an endless vista of undulating yellow grasslands, smudged just here and there by the grey–green wash of eucalypts and, more recently, by darker, isolated splashes where planted pines and poplars hid homesteads. This little greenery from England and the homestead gardens were the only concessions allowed by the dry, hard land. Inland Australians had long given up trying to impose traditional beauty from a different land. Finding the landscape to be uncompromising and against change, they eventually awakened to its beauty.

Time was measured by the seasons. Once the hardships of settlement had been overcome it marched to the casual pace and easy laughter of Saturday afternoon tennis parties held in the cool of the homestead trees, and to the changes in the appearance of children whose brief visits marked the rotations of boarding school terms.

Progress was measured by the annual Pasture and Agricultural shows, where achievement could be a blue ribbon for a fine looking ram, or first prize for a jar of home-made marmalade. Like the bogong corroborees which every year had attracted far-flung Aboriginal tribes, the agricultural shows brought together the widely scattered European Australians, who danced and courted, exchanged gossip and news, and listened to brass bands whose sweet harmonies evoked memories of far-away England.

Isolation had welded the people into small, independent communities. They were of resolute pioneering stock, living in homes fashioned by their forefathers from the materials at hand. They identified emotionally with the exuberance and daring of young stockmen who made their start by capturing mounts from wild brumby mobs and built up their herds from rogue cattle that had gone missing from past musters. It was a time when the jingle of bit and spur, the snort of a horse and the creak of leather were still familiar sounds as dawn began to light the dewy plains.

Every summer, like their fathers and their fathers' fathers, they drove their sheep and cattle into the mountains to rest home pastures. The stock thrived on the mountains' natural clovers and snow flowers. Nothing changed. In 1949, as in 1849, stockmen would gather at the great runs for the summer droving and the autumn musters. Every rider was a master horseman, always eager to demonstrate his ability and courage. They were resilient bushmen, fed on slabs of damper, cold beef, strong black tea and campfire yarns; the sons and grandsons of the men from the Snowy River who rode to join the famed Australian Light Horse in the imperial wars against the Boers and the Germans. The men of the Monaro 7th Light, on mountain-bred horses, could ride, and fight, like few others.

The small townships which served and fed off the graziers mirrored their ways and shared their heritage. Even in the early 1950s, stockmen were still a common sight, droving sheep through large towns such as Cooma as they made their way slowly, noisily and dustily to the saleyards. Friday was sale day. It always had been and there was no reason yet to think it would change.

The movement of the grazier gave rhythm to the town's life. It flowed as a tide to his comings and goings, his selling and buying. The graziers and their families always stayed in town on Friday night for the Saturday ritual of collecting groceries and farm supplies and joining other graziers—and the bank manager—at Murphy's store.

Murphy's, on the corner of Sharp Street and Vale Street, was one of those small-town institutions. The men gathered in the cool depths of its cellar to yarn with friends from across the plain and share a tot or two of 'the stronger stuff'. Bullock's whisky and overproof rum were the drinks of the day. Beer was rarely drunk, except over the bar in a hotel. It was too bulky to carry away on a horse.

The strength of the bonhomie generated at Murphy's could be measured on Sunday morning by the parcels of groceries and provisions that still sat outside on the sidewalk awaiting collection. Theft was unheard of.

Murphy's was run by two local identities, Norman and Stan Dykes. Stan was local buckjump champion. During World War II he organised the Snowy cattlemen into a touring rodeo troupe which, between 1941 and 1946, raised 20 000 pounds for the 'war comforts fund'. The 'old-timers' still talk about the dust they raised in Sydney with local horses like Hell's Angel, on which no rider ever survived longer than six seconds.

But for many the war meant much more than fundraising rodeos. Like their fathers twenty-five years earlier, many left the high country to rally to the cause of an England they had never known, except from schoolbook history. Now, in the graveyards of even the tiniest towns, there were memorial headstones, some still shiny and new, detailing in chiselled granite the heartache of a loss on foreign soil.

Beneath the whispering pines that sheltered the old cemetery above Adaminaby, headstones in neat rows separated and chronicled the generations stretching back to the birth of the settlement. It was a continuum.

But there was something unsettling about the newer epitaphs on the hill high above town whose haunting Aboriginal name, *Adamindumee*, means the 'resting place':

Private Edmund Joseph Baragry
Killed in action, Bullecourt, France
9th May 1917
Aged 22

JINGLE OF BIT, CREAK OF LEATHER

Old Adaminaby c.1951.

—a boy from the high country, who responded to the frenzied rally of a dying empire. Sixty thousand of the country's most able-bodied young men were killed in that war which lopped the population and set back Australia's development as an independent nation by half a century.

Twenty years later there was more tragedy for the stonemason to chronicle:

Ross Mansfield
died
POW Burma
5th October 1943
Aged 23

He was one of many young men from the high country, sent with his mates to another war—which this time came perilously close to home—to protect and secure the future of their friendly, green valley. They were never to know that so soon after their sacrifice the valley would itself die, drowned in a man-made sea.

Now this young Italian's grave stands in the overgrown grass of the Old Adaminaby cemetery among the sun-bleached headstones of Australian pioneers and their descendants.

In Memory of
NARDINO PELLEGRINO
Aged 30
Born in Piticliano, Italy
Remembered by all his friends

He was killed in a truck accident while building the very dam that would inundate their valley.

The potential of the snow thaw had been known for long enough, but those who spoke of it were 'dreamers'. To the people who knew the mountains, the size of the engineering project that had been mooted from time to time seemed an impossibility. When word filtered through that the government was seriously considering a scheme to harness the alpine waters, the debate on the Monaro was confined to bush logic. Nobody really knew what it would mean and so they limited their views to simple statements of optimism or ridicule.

Rumours flew from one end of the Monaro to the other. There developed an atmosphere of tension mixed with anxiety and sadness, for change would inevitably bring its share of heartache.

But even as the Monaro people argued, there were strangers in their midst—high up in the mountains with theodolite and pole, marking the earth and rock for the invasion to come. In fact, within weeks of the decision to go ahead, there were men in the mountains, exploring the deepest ravines and steepest ridges, out of sight of the communities on the Monaro. These men from foreign lands gazed wide-eyed at their surroundings as they followed the instructions of a wiry old man, white-haired under

Milisov (Jacky) Tsarevich second from the left at the surveyors' camp, Indi, 1955. Second from the right is Stan Kajpust and far right is Kon Martynow who worked with Major Clews on the start of the Tumut project.

a battered 'mounties' hat. Just as the high country people had yet to meet these strangers in their territory, the newcomers had no concept of the community they had entered.

A few months earlier, Milisov Tsarevich had been working his family's small, mixed farm—a random collection of chickens, pigs, cows, a few sheep and lambs—about a hundred kilometres north of Belgrade in Yugoslavia. Life there had been a struggle. As the eldest son he had shouldered the responsibility for keeping the family and farm together while his father and uncle were in German prisoner-of-war camps. It was a harsh existence, and the peace which should have relieved it was no less oppressive. To a young man with far horizons in view, the new regime represented constraint and rigidity of thought. While the vast majority acquiesced, some, like Milisov, found freedom an all-consuming desire. He put his faith in God, challenged the border, and went searching for a new life and a new home. In 1949 the refugee pipeline deposited him in Australia.

While the clientele at Murphy's were waxing lyrical over the Snowy Scheme's intrusion into their lives, Milisov Tsarevich was already there, a foreigner working deep within their alpine domain. Beside him was another—Dusan Sbremac, a Serbian surveyor. Apart from immigration officials in Melbourne, which seemed a world away, neither of these young men had had more than fleeting contact with the local population. It all seemed surreal. Nothing in this new world was familiar; nothing related to their past experience. Perhaps it was all part of a dream they were sharing—bush cooking on an open fire, the exotic smell of eucalypts, the strange animals they had no names for. And yet they felt that they belonged. If they had any fears, they were not of the dream but only that the dream might end.

The bush, in which a person could become lost in a matter of seconds, frightened even the bravest. But it is understanding, not courage, which makes it a friend. For a chosen few, the mountains wove their ancient spell. Milisov, the farm boy from beyond Belgrade, was one of them. He adjusted to its moods—its melancholy greens embraced by thundery skies; its radiant valleys touched with afternoon gold; the orchestra of the night and the silence of the dawn. Milisov saw nothing in the bush but lavish and breathtaking splendour:

> I loved it as soon as I was in it. It was wonderful...a bit cold and damp but as a young man I didn't mind. I'd see possums, wombats and cockatoos. I didn't know their names but I watched them and learned to understand them. The cockatoos and currawongs would warn us when to secure our camp, telling us that the wind and rain would soon be on us.

The bush infused itself into Milisov's blood. He went once to Sydney but was back just four days later. The Snowy became his home, the place where he would live for almost the rest of his working life before he settled with his Australian wife and family in Tumut, at the foot of the mountains.

For some, like Milisov, it was several years before the stranger in the mountains and the Australian on the plain met face to face. But by then, the man who had arrived shy, gaunt and hollow-eyed would be gone; in his place a bushman with a knowledge and love of the country as profound as that of any high country stockman. The meetings were friendly with, in Milisov's case, the typical Australian response to the introductions: 'Mind if we call you Jacky, mate?' And Jacky it is to this day.

At other times the first contact had a startling effect. The German workers at Wambrook, one of the early roadworks camps established to build a new Cooma-Adaminaby road, were excited by the large numbers of rabbits to be had for both sport and the dinner plate. One Saturday morning a group of fifty or so marched into Cooma and bought out the town's entire stock of shotguns and rifles. Word spread like wildfire that the Germans were arming themselves. There was near panic as doors and windows were bolted and barred and the people prepared for the outbreak of World War III.

For others the whole thing was just very confusing. Sixteen-year-old Boyd Mould was staring with a stolid expression out through the carriage window as the train carried him home from school holiday in Sydney. Normally he would have had the carriage to himself. Now it was full of strange men who smelled absolutely awful. Only by fixing his gaze on the sliding landscape could he hold on to his identity, confirm that he was in his own time and place. As the hills out of Canberra flattened into the Monaro he breathed easier, but still through pursed lips. It would be some time before he recognised the pungent odours as garlic and spiced sausages.

Boyd had heard of the Snowy Scheme, but only through oblique discussions between his father and neighbours about whether it would affect their mountain grazing leases.

Boyd hoped it would not. He fully intended to be a grazier in his own right when he left school. The thought of losing what they all felt was theirs by right was intolerable. He could not comprehend a world not bound to the alpine musters. Besides, how did the government expect them to be able to maintain viable stocking rates if they could not use the Alps to rest their pastures? What would happen to the stockmen and characters like old Charley Spencer, whose grandfather, James Spencer, was accorded the honour of being the first man to take stock into the mountains, way back...well, it was more than a hundred years ago, he knew that.

Besides, old Charley was a bit of a legend himself. God, he made them laugh! Old Charley had a withered arm, smashed after a fall on a horse. Now when he rode he wore a coat so that he could tuck the useless hand into the waist pocket. But when he galloped, it always came out, flapping uncontrollably and often flying up to hit him in the face. The boys laughed all right, fit to bust. Boyd grinned through the dusty glass.

The stockmen were the people who belonged on the plains, not these strangers who couldn't even speak English. He remembered the story his dad, Reg, told about a night he camped with old Charley, and chuckled quietly to himself despite his uneasiness.

As was often the case, the campfire talk got around to dogs: exploits of the best working dogs they had owned and the tribulations they had suffered at the antics of the more useless mongrels. Charley was poking the fire with a stick, sending a fine spray of sparks into the night. 'Actually Reg, I'd like you to come around one day and have a look at a young pup I've got. I want to know if you think he'll be any good as a worker.' Boyd's father was modest. He gazed thoughtfully at the flames as they licked the dead branches in the centre of their fire.

'Well, I'm not really an authority on these things, Charley.'

'Course you are. You've always got a good dog. Come on, come around and tell me what you think of 'im.' Reg reluctantly agreed, and a week or so later called in to give Charley his verdict. Charley waited anxiously.

'Well, what do you think?'

Reg shrugged: 'Well, he's a friendly little fellow all right. You should be able to teach him something—plenty of width between the ears; shows he's got some sort of a brain.'

Charley spat contemptuously: 'Ah strewth, that buggers it then. He'll never work if he's got any brains.'

The story was told and retold around at least one campfire on every muster. Everybody knew old Charley and laughed. Funny as a headless chook, but a good bloke, was old Charley.

These were the characters from the familiar and secure world he knew. Boyd made his first trip into the mountains when he was four and a half years old, sitting in front of his dad on a big brown horse. He still remembered it in timeless snapshots pasted in the back of his memory—particularly the fierce winds and burnt leaves. There had been a big fire which had killed seventy people.

He had gone with his father on almost every muster since, and was looking forward to the summer holidays and the next big stock drive up to their lease. With a good dog, he and his father could comfortably handle a mob of up to 3000 sheep. With a good dog he could take 2000 on his own.

School meant he missed the autumn round-up to bring the stock back home, but the May holidays gave him the chance to go back with his dad to gather in stragglers before they got marooned in the snow. It was usually a six-day trip: horseback by day and a campfire by night. Most of the stockmen had been in the mountains all their lives. They all had stories to fire a boy's imagination.

The stock route was from the farm near Dalgety, to Berridale, from there to Rocky Plains, then the Eucumbene River near Nimmo, climbing to Snowy Plains, and from there to wherever the lease was. The droving took them through ever-changing vistas—from the tea-trees that lined the creeks, into the mountain ash, and finally up among the snow gums. No matter how often they went into the mountains they never tired of their beauty; and never dropped their guard against their capacity to unleash from nowhere a blizzard that could trap man and beast for days—particularly in the early summer and late autumn.

The snow could be as heavy as it was sudden. One winter, Kiandra postman Bill Paterick recorded two and a half metres of snow in eight hours, a rate measured then as one foot every hour. In such conditions the landscape was transfigured overnight.

Bill had the mail contract, delivering to the remote stockmen's huts and miners' shacks. In summer he travelled on horseback and in winter on skis fashioned from planks of timber. The steep, winding dirt road that linked Kiandra with Adaminaby, the nearest town, was snowbound for up to six months of the year.

In 1949 Kiandra had, apart from the post office, a butcher shop, general store and a one-classroom school which served a permanent population of about thirty. The numbers swelled to a hundred or so in winter when skiing enthusiasts made the long trek into the mountains and bunked down in the Kiandra chalet. The skiers provided the only real contact with people from the world beyond the mountains.

The Kiandra folk had learned about the possibility of the Snowy Scheme the previous year but, as in Cooma, it did not mean much—except perhaps to Bill Paterick's son, Max, who had mixed feelings about spending the rest of his life minding sheep or burrowing for gold.

For Max life as a boy had been wonderful—skiing to school, fishing in the Eucumbene river, horseriding through the mountains and joining up with mustering teams. But sometimes he and his two brothers had to help their father panning or working his mine. It was hard work pushing the laden skips in the small, confined tunnel—especially when they went six or seven hundred metres into the mountainside, where lack of air made it difficult even to breathe. Max had now just turned seventeen and did not want to do this kind of work for the rest of his life. In any case this was hardly likely as most of the accessible gold had already gone.

Then came the Snowy, and Max was handed an unexpected alternative. The Scheme was just a few weeks old when he won a contract to supply horses for the surveyors. He was effectively a one-man support team: 'I used to take the surveyors out and build their camps for them. I knew every inch of the country from Yass to Victoria. I knew where the water was and where the good campsites were.' Max was one of the first locals to have contact with Major Clews and his migrant survey teams but he chose mainly to ignore the foreigners: 'I couldn't understand them and they couldn't understand me, so I just did what I had to do and left them alone.' When the surveyors had gone, Max did the same work for the heavily burdened diamond drilling teams who followed in their footsteps. They drilled for rock core samples in the search for suitable foundations for the giant dams.

Max was typical of the young high-country men. He was raised in a world where, from boyhood, horsemanship and skiing had to be second nature—a world that the rest of Australia, far away beneath the valley mists, hardly knew existed. Max's intimate knowledge of the mountains and his skill as a horseman and skier later saved the lives of twenty-five dam builders who had been stranded for ten days in a blizzard.

The stockman of the high country had to be able to read the weather as a motorist reads a roadsign: to know in advance when bad weather was on the way. He also needed to know intimately every lease he was despatched to mind; to be aware of what lay beneath each new blanket of snow, the location of hidden crevices, deep drifts and creeks which could swallow a horseman or his stock. The saving of a herd or flock, often after a lone and exhausting battle against the fury of a blizzard, made a man a legend among his kind. But it was a matter of survival, not heroics.

'The sheep were our bank, our only source of income. If they perished, we perished.' That is how Herb Hain describes his years as a grazier on the Monaro. 'You had to watch the weather daily, sometimes hourly, looking for signs of an early autumn or late spring. You needed that knowledge to keep yourself and your stock alive.'

In the autumn of 1944 it took him fifteen days to rescue and drive to safety a vast mob of more than 100 000 sheep, trapped and starving near Kiandra. The snow had come early and the first falls were so heavy that, even down on the plain, the Cooma to Nimmitabel train became stranded. Herb rode into the mountains with two other stockmen, his neighbour, Harry McGufficke and good mate, Cliff Rose. At the lease they were joined by another stockman, Jimmy Patterson, who had skied across the top of the mountains from Kiandra.

They found the sheep huddled in small groups inside deep pits of snow which had built up around them. The men spent eight days digging into the snow walls to build ramps that would allow the animals to escape onto the frozen snow surface. Here they could feed off protruding leaves. Suffering from snow blindness and exposure, the men sheltered in a hut for another week, living on salted beef, potatoes and damper. Then, when they had recovered and the weather had eased, they mustered the sheep into a large mob and herded them to safety down onto the plains.

The weather was an accepted risk in stock work but that didn't mean you were never worried or scared. To get caught in the snow was pretty rough. If a storm hit, you had to get out as quickly as possible; leave the stock and come back for them later. If you were on foot or horseback, and you didn't get out before the snow froze, there was a good chance you would never get out. A sudden blizzard is a frightening experience, no matter how many times you've faced it.

For men working alone it was doubly dangerous. In the winter of 1937 a stockman named Jack Adams, a hardened old-timer of the high country, fell from his horse and tumbled down a rocky slope, breaking a leg and suffering deep cuts to his body. In agony, but driven by the knowledge that he would die lying where he had fallen, he crawled back up over the rocks to where his horse waited. With a barely imaginable effort he hauled himself onto his saddle and let the horse carry him to his hut. Sliding off the mount onto his good leg, he dragged himself inside and collapsed onto his bunk.

Early the following day Herb Hain rode by, looking for spare stock horses left behind during the autumn muster. He saw Jack's horse outside the hut and wondered why the man had not appeared to wave and yell a cheerio. He noticed the horse was saddled, assumed the man was busy and about to ride out; so he continued past. On returning later that day, with his mustered horses roped together in file behind him, he saw Jack's mount still saddled and untethered in the same place; just standing restlessly outside the hut door.

Sensing something was wrong, Herb rode up. Inside, he found the stockman chilled and almost lifeless. The man's body and bedding were covered in congealed blood and a thin cry of pain came in waves through his cracked lips. With the knowledge and calm of an experienced bushman, Herb worked fast, lighting a fire and gently forcing a lukewarm mixture of rum and water into the injured man. He then tethered the spare horses and set off through the snow on his own mount to the closest settlement, the Charlotte's Pass chalet. It was a long, frustrating ride. Where the snow was thin or had frozen hard, his horse could manage nothing faster than a brisk trot. At the chalet a rescue team was assembled and the injured stockman was eventually carried out and taken to the nearest bush nurse, at a homestead on the lower slopes. Jack Adams was regarded as one of the most experienced stockmen in the mountains; but after his accident he never returned. He simply disappeared from the world of the high country.

Herb Hain knew and loved the mountains as deeply as anybody. His great-grandfather settled on the Monaro in 1854. His family owned some of the largest acreages in the region, as well as two hotels and a store in Cooma. But Herb knew change was in the wind; and with equal conviction believed it would be for the better. The determination of men to dam the alpine waters held no fears for him. He had long felt that the Snowy was destined one day to become a resource of great value to the nation.

When the Snowy Mountains Hydro-Electric Authority arrived finally to give concrete form to these ideas, Herb Hain was quick to respond. He used his knowledge of the mountains and his previous experience—during the Depression he had opened a butcher shop in Cooma to sell farm meat he could not quit on the Sydney market—to win a contract to supply meat to the work camps.

In winter the rough access tracks cut by the Authority turned into axle-deep mud, but Herb's supplies usually got through. He exchanged his horse for a jeep and trailer, but even then he still occasionally got caught. He once spent a day and a night trapped with a load of pig carcasses: 'I had an overcoat, some sandwiches and a thermos and just had to sit it out. It was so bloody cold there was no worry about the meat going off.'

Herb leased, and ultimately sold, his butcher shop to a group of enterprising migrant workers who saw a captive market for smallgoods. The operation later changed hands again. This time it was bought by two brothers, the Handler brothers, who became for a while one of the largest smallgoods manufacturers in Australia.

The Snowy River flowing through Jindabyne, 1950.

When the government finally enacted legislation to begin construction, the Cooma business community heartily welcomed such a project to their region. But they were totally unprepared for the social upheaval it would bring. And few were prepared for the fate that awaited nearby Adaminaby and the town and beautiful valley of Jindabyne. Only the graziers seemed to understand. If this Snowy scheme was as big as some were saying, it couldn't help but affect their lives. But they, too, were only half prepared for what was to come.

Cooma, with a population of about two and a half thousand would, in just three to four years, explode. Its population would climb to over 10 000 and the locals would be completely overwhelmed by outsiders who would bring with them strange cultures and foods, twenty-four-hour nightclubs, restaurants, illegal casinos and brothels. After a gunfight in its main street, Cooma attracted headlines describing it as a lawless frontier town.

The Snowy Mountains Hydro-Electric Power Act was enacted on 7 July 1949 and the newly formed Snowy Mountains Hydro-Electric Authority was quick to start work. On 17 October, in a small ceremony at a spot near Adaminaby, a flag was nailed to a mast to mark the start of work in the area (eventually more than 3200 square kilometres) to be occupied by the Scheme. The Governor-General, Sir William McKell, who just a few years earlier as the Premier of New South Wales had fiercely opposed the electricity generating function of the Scheme, detonated an explosive charge at the site of the proposed Adaminaby Dam and declared the work begun.

The ceremony was a reassuring event but it rather disguised the true state of play.

The Commonwealth and States Snowy River Committee had drawn up the Scheme only in sufficient detail to justify its adoption. The detailed geology, hydrology and survey work, necessary before the Authority's engineers could begin planning a construction program, had yet to be done. In order to avoid delays that might stall initial momentum and strengthen opposition to the Scheme, this work had to begin immediately.

A general election was due in December and the signs were that the Liberal–National Party coalition under Menzies would topple the Chifley government. There was increasing public antagonism over the coal strike and the government's proposal to nationalise the banks, and it had already clearly lost the rural vote with the effects of its taxation reforms on rising wool incomes.

The impact of a change of government on the Snowy Scheme could be immense. There was considerable doubt about the level of support Menzies would give, and even about whether he would sanction its continuation. Liberal politicians had pointedly boycotted the opening ceremony at Adaminaby. Menzies was regarded by some in the political arena as a man brilliant with words but shallow—all gloss on the surface but lacking substance; a masterful politician, but one whom the Snowy people did not yet trust.

Opposition also remained strong and bitter from elements in the New South Wales Labor government, still smarting at losing the argument for the simpler irrigation priority. These elements were engaging in a brazen but determined campaign of disinformation to discredit the decision to divert water into the Murray River for power generation. Joe Cahill, the New South Wales minister on the Commonwealth and States Snowy River Committee—later to become Premier—bluntly confronted the federal Minister for Works, Nelson Lemmon. 'We're doing everything we can to kill this scheme of yours. I'm sick of coming to Canberra to listen to your bloody pipedreams,' he told Lemmon. 'This scheme won't work and we're going to stop it any way we can.'

Cahill protested long and hard that the Commonwealth was interfering with the sovereign rights of the states. But there were three other reasons why New South Wales did not want the Snowy Scheme as the Commonwealth envisaged it: it favoured its cheaper irrigation-only option, with consequent benefits flowing almost entirely to New South Wales; it had expansion plans for its own state electricity utility using coal; and it believed the Commonwealth's scheme posed a massive engineering problem that was beyond Australia's capabilities. On the latter point, Lemmon responded that if they could build one dam and one power station, then why not twelve dams and twelve power stations? To him it was only a matter of repetition and scale.

But Cahill would have none of the argument and set about sabotaging the plan, using agitators and a campaign of disinformation. He was particularly active in the Riverina district which was to have been the main recipient of the waters under the New South Wales proposal. He enlisted an agitator named Gleeson from the Riverina town of Leeton. Letters circulated by Gleeson became so extreme that Lemmon eventually confronted the man and accused him of being Cahill's stooge—a charge he did not deny. Lemmon persuaded Gleeson to meet one of the Scheme's senior engineers. On seeing for himself what was proposed, and realising that water was still to be diverted to the Murrumbidgee for irrigation, Gleeson publicly dropped his opposition to the Scheme.

But opposition from graziers who were losing their alpine leases had begun to increase.

Prime Minister Ben Chifley, Governor-General Sir William McKell, and Minister for Works Nelson Lemmon on the Union Jack draped dais to mark the commencement of the Snowy Mountains Scheme at Adaminaby, 17 October 1949. The ceremony was boycotted by the Liberal Party.

PHOTO: NATIONAL LIBRARY CANBERRA

The graziers did not believe there was enough rain in the mountains to supply the Scheme reliably, and felt they were being pushed out to make way for a lavish white elephant.

At a meeting organised by the Snow Lessees' Association, the Authority presented graziers with a facts-and-figures profile of the Scheme. The graziers were told that Lake Eucumbene would take twenty years to fill and that half its capacity was in its top one and a half metres. This sounded alarm bells among the graziers whose experience was that farm dams in the area lost between one and a half and two metres in evaporation. So if Lake Eucumbene took twenty years to fill, it would be half empty a year later, even if little or no water was taken to feed the Scheme. The Authority dismissed the argument—though in 1988, after a prolonged dry period, the total water reserves of the Scheme fell to as low as forty per cent of capacity.

With such pockets of intense opposition and uncertain political support, everybody involved with the Scheme during its early years felt the pressure of time. They were driven to establish it as a fait accompli as soon as possible.

The Authority urgently started a systematic mapping of the mountains' rock formations to obtain the information about the nature and distribution of rock and soils that it needed for the planning of dams and tunnels. Detailed work also had to be done by hydrologists. According to their size, altitude and terrain characteristics, the catchment areas varied widely in their contribution to the annual run-off. Such information was as important as the geological structure for the siting of dams. Because of the shortage of surveyors in the 1940s and the inaccessible nature of much of the area, the government also drew on the services of the Army Survey Directorate. Its surveyors helped with the work being done under Bert Eggeling and the 'retired' Major Clews.

On occasion, Nelson Lemmon went by foot and packhorse with the surveyors—not only because of his interest in the ground-level details of the Scheme, but because he was conscious of the need for environmental protection, long before it became a political issue. Lemmon urged the surveyors never to take a cheap and easy short cut if it meant scarring the landscape. He foresaw that tourism would follow the Scheme and that people would want to see the country as it was.

As the Scheme developed, its impact on the environment inevitably drew some criticism. But, despite what some have suggested, the Scheme's architects strove hard to minimise environmental damage and placed a high priority on land restoration. Before the Scheme even began, the fragility of the alpine environment was understood; it was the reason that the highest grazing leases had already been cancelled. By 1955 the Authority's budget for soil conservation alone was 200 000 pounds, a vast sum in those days. Stringent environmental obligations were built into the Act governing the Scheme.

As the geologists, hydrologists, surveyors and other specialists completed their field and laboratory tasks, the information they supplied was correlated and studied by those responsible for the overall shaping of the Scheme. In a continuous and accelerating process, their decisions were translated into designs and designs into structures.

In an effort to appease New South Wales' irrigation interests along the Murrumbidgee, the Authority decided that the Tumut system, using waters from the northern catchments, would take precedence in the time schedule over the Murray system.

The Snowy-Tumut Development, as it was titled, incorporated the diversion of the

Eucumbene, the upper Murrumbidgee and the Tooma rivers to the Tumut River. In their 800-metre fall before release to the Tumut, and thence the Murrumbidgee, the combined waters of these rivers would generate electricity in four projects: Tumut 1, Tumut 2, Tumut 3 and Blowering.

The development necessitated the construction of a trans-mountain tunnel system. The main project was the Eucumbene–Tumut tunnel, to connect the proposed reservoir on the Eucumbene River (today known as Lake Eucumbene) with the Tumut Pond reservoir which was to be built on the upper Tumut River. The normal function of this tunnel would be to divert water westwards through the Great Dividing Range from Lake Eucumbene to the Tumut River. But during periods of high flow in the Tumut and Tooma rivers, excess water could be pumped back for storage in Lake Eucumbene.

But the size of the project, and the exhaustive groundwork to be done before construction could begin in earnest, meant there would be a considerable time lag before the Authority had anything tangible with which to justify its existence. The Snowy–Tumut Development was expected to take nine to ten years to complete.

As a matter of urgency, therefore, the Authority decided also to begin work on a small hydro-electric project which was independent of the two main developments and which could be completed relatively quickly. This was to consist of a power station on the Snowy River, twelve kilometres from the summit of Mount Kosciusko. It was to use the water of the upper Snowy River, dammed at the point where the Guthega River joined it. The station could be in use within five years and would demonstrate the Scheme's value by feeding power into the New South Wales scheme while the major works were still in progress.

However, little of this exhaustive planning was public knowledge. Most Australians at this stage were only vaguely aware that something was going to be built in the Snowy Mountains. And whatever it was, it was too remote to arouse more than a mild curiosity. In Murphy's cellar, when the talk turned to the Scheme, the discussion, after some initial excitement, tended to become academic. There was no immediate activity that even the locals could see.

In the beginning, the Authority's only visible presence on the Monaro was at Cooma where several rows of tents appeared one afternoon on open, undulating land that had been acquired on the northern outskirts of the town. They reminded the locals of the early goldrush days at Kiandra up in the mountains. The tents were not a threatening presence and for a short while they eased the disquiet some felt about the Scheme.

The tents, however, were just the start of what the commissioner, Hudson, intended to be a whole new community. It would soon have houses, rows of barracks, streets and sewers—the likes of which Kiandra had never known, and which even in Cooma represented a considerable step forward. The Authority, with its teams of engineers and scientists, was about to drag the whole district into the twentieth century.

At first, the Authority's headquarters was in Sydney. A single room in a commonwealth office grew to two rooms, and before long the Authority occupied a whole building in O'Riordan Street in the suburb of Alexandria. There, Hudson was gathering together his initial management team. They were men whose names would become worldwide in the field of construction and hydro-electricity.

Among the first engineers appointed were E.F. Rowntree and Olaf Olsen. Olsen was an experienced Norwegian engineer who was previously employed by the State Electricity Commission of Victoria. It was Rowntree who had developed one of the fundamental concepts of the Snowy Scheme—the diversion of water to the Tumut River for power production and irrigation. And it was Olsen who first argued the possibility of diverting water to the Murray for power production, an argument that had given the Victorian and commonwealth governments their answer to New South Wales' demands that Snowy water be diverted into the Murrumbidgee exclusively for irrigation.

Others who joined the Authority in the first year and were appointed to, or ultimately attained, senior positions were Ken Andrews, chief engineer, major contracts; Bob Archer, secretary to the Authority; John Daffy, business manager; Joe Hinton, chief administrative officer; Arthur Humphrey, chief finance officer; Tom Lewis, field construction engineer; John 'Darby' Munro, field construction engineer; Iver Pinkerton, engineer in charge, civil engineering design; Albert Ronalds, chief civil design engineer; Ian Sergeant, engineer in charge, investigations; Darcy Walsh, engineer in charge, field investigations; John Waldren, superintendent of stores.

The government also appointed two associate commissioners: Thomas Arthur Lang,

Olaf Olsen, the Norwegian engineer who first proposed diversion of the Snowy River to the Murray River for power generation.

previously the commissioner of irrigation and water supply in Queensland; and Tony Merrigan, who had been senior electrical engineer in the Victorian State Electricity Commission. Lang was a civil engineering graduate from the University of Melbourne and before his Queensland appointment had been with the Victorian State Rivers and Water Supply Commission, ultimately as its chief design engineer. Merrigan, an electrical engineer, had also graduated from the University of Melbourne. He had held engineering positions in the Victorian State Electricity Commission, General Electric and the Brooklyn-Edison companies in the United States and the Hydro-Electric Power Commission of Ontario, Canada.

In December 1949, when the Scheme was barely six months old, the Chifley government did fall and doubts about the Scheme's constitutional validity, raised by Menzies when he was in Opposition, attained renewed prominence. A judgement in the High Court, declaring the Scheme unconstitutional because of the way the federal government took control of territory which had previously belonged to New South Wales, would provide the government with an easy political 'out'. If there was a successful court challenge by some interested party, the consequent collapse of the Scheme could be blamed on the appellant, with the government seemingly remote from the case and beyond reproach. Alternatively, Menzies could simply starve the Scheme of funds until it withered and died, and could easily be buried as a brief, foolhardy idea.

The first post-election statement of support for the Scheme from the government did not come until the first session of Parliament in 1950 when the member for Farrer, David Fairbairn, said in his maiden speech that the Scheme would greatly benefit his electorate (the Albury district on the banks of the Murray) and he hoped every effort would be made to speed its construction. When the Budget estimates were presented in October, they included twelve million pounds for the Scheme; but still its future seemed clouded. The government was becoming increasingly concerned about the prospect of another war; and with the soaring inflation which was threatening the country's economy. In March 1951, the Minister for National Development, Richard Gardiner Casey, announced the Scheme would have to stand up to some critical examination if it was to continue.

In the same month Prime Minister Menzies told the House of Representatives that he believed the state of the world was such that Australia had no more than three years in which to get ready to defend itself. Menzies sometimes sounded as though he begrudged not having had the opportunity to be a wartime Prime Minister. In Parliament he named factors pointing to another war as the weakness of Western European defence, the extent of Communist military resources, the disturbing position in the Middle East (which later flared into the Suez crisis) and Communist activities in Malaya, Indo-China, Korea— and Australia.

A few weeks after the Casey and Menzies statements, with the future of his grand dream still hanging by a political thread, Ben Chifley, the visionary former Prime Minister who had so enthusiastically supported the Scheme, died.

The Scheme did finally get the support of the new government, though it would be some years before the state governments, particularly New South Wales, sanctioned the plan. Until then, every move the Authority made had to run a bruising political gauntlet.

CHAPTER FIVE

The first Snowy camp—for road workers—on the outskirts of Jindabyne, 1950.

CONVERGENCE

CONVERGENCE

The world was wet and miserable when Jock Wilson stood on Cooma railway platform and stared forlornly into the mist that shrouded the town. His spirits were low. If the night train which had brought him from Sydney to this saturated corner of the earth had not been so cold and uncomfortable, he would have climbed back on board. Instead he joined a line of equally grim, unshaven men who were filing desultorily into an ancient mud-streaked bus.

The bus took them on a short, jolting journey along a puddle-pitted road before hitting the town's only strip of bitumen—its main street, Sharpe Street—where it deposited them at a cafe opposite the Alpine Hotel. The hot toast and tea—courtesy of their new employer, the Snowy Mountains Hydro-Electric Authority—was welcome solace from the all-pervading gloom. To be a stranger under a thundery, wet sky was a disheartening experience.

Warmed a little by the food, the men were herded back aboard the bus and taken to the Authority's office in Vale Street. As the bus braked convulsively to a stop its door was heaved open from the outside and a weathered face poked through, deep-set eyes quickly taking in the faces which stared back. The man, Dan Murphy, saw it all: homesickness, uncertainty, confusion—emotions he was becoming adept at both reading and treating. He grinned wolfishly: 'Right-oh, you bloody new Australians, out.'

He jerked his thumb in the direction of an open truck parked nearby: 'You're going to Jindabyne.' The men began to shuffle out, heads bent under the low bus roof. They had understood the language of his thumb. Jock Wilson turned to a New Zealander he had befriended. 'Does that include us?' he asked in his broad Scots burr.

Jock had been in Australia for six weeks. In that time he had already drunk billy tea, chewed damper and slept under the Southern Cross. Compared with the 'foreigners' on the bus he felt that should already make him an Aussie. Jock had come to Australia to work as a jackaroo for a prominent South Australian grazier, 'a bloke named McLachlan'. But there was a disagreement over the employment conditions, so after three weeks Jock quit and went to Sydney where he learned of workers being recruited for the Snowy Scheme. He was among the first to sign on at the Sydney office before the Authority moved its headquarters to Cooma.

The New Zealander, too, was unsure about the precise definition of a 'new' Australian, so both men stayed put. After the others had filed from the bus, Dan Murphy put his head back through the door, gave the pair a brief quizzical glance, then withdrew, pulling the door shut as he went. The bus driver turned around to face them: 'So you're going to Three Mile then, eh?' Jock turned to the Kiwi: 'What's Three Mile?' The New Zealander didn't know, but they both nodded an affirmative to the driver. Anything was better than the alternative open truck ride in this weather. The driver yelled above a screech of metal as he forced the gearbox to accept his will: 'Right. It's a bit of a trip, and I've got to pick up provisions. Don't mind, do you?' The pair indicated nonchalantly that of course they didn't.

The first work camps for the Snowy Mountains Scheme were in three areas: on the outskirts of Cooma where the Authority planned to establish a permanent headquarters; and near Jindabyne and Kiandra, where road building crews were based for the initial thrust into the mountains. Three Mile was the camp near Kiandra, named after a dam built late last century by Chinese gold miners to provide a permanent water supply for sluicing.

It was at these camps that migrants, mainly Yugoslavs and Germans, first appeared. But their numbers during the early months were small, and most of the newcomers kept to themselves. They were kept very busy. The decision to push ahead as fast as possible with the independent Guthega project on the Snowy River below Mount Kosciusko necessitated an urgent road-building program.

The first task was to upgrade the track to the Kosciusko Hotel and Charlotte's Pass chalet. From this road, at a place called Digger's Creek, a new road was to be pushed northwards to Island Bend which was destined to become one of the more famous sites in the saga of the Snowy. From Island Bend the road would turn sharply to the southwest and climb even higher to the site for the Guthega power station and dam. At the same time, the Authority started bulldozing a steep, twisting descent from Kiandra, via Three Mile Dam, to Tumut Pond. This flat on a bend in the Tumut River was considered an ideal site for the main work camp for the construction of Tumut Pond Dam.

From Cooma, it was a long, slow journey over rough dirt roads. The bus driver taking Jock Wilson and his new-found New Zealand friend to Three Mile camp had not been exaggerating when he said it would be 'a bit of a trip'. The roads were a quagmire. It took them all morning to reach Adaminaby, at the foot of the mountains fifty kilometres to the north-west of Cooma. There they stopped for lunch before a journey, halted by frequent breakdowns, up the narrow, twisting road which climbed the face of the mountains before crossing the eastern ridge and curling westwards along a plateau to Kiandra. It was late afternoon when they reached the outskirts of the tiny mining and skiing settlement on a rolling treeless plain. Granite mullock heaps marked where men had sought their fortunes by burrowing into the very rock of their promised land. In its early days nuggets as big as twelve kilograms were found, but the alluvial gold quickly petered out, forcing men to tunnel in search of seams. Few succeeded, and now just a handful of old prospectors still toiled—a far cry from the goldrush days sixty years earlier when thousands of men braved blizzards, isolation and deprivation in the hope of finding just a few grains of the precious metal.

At Kiandra the bus and its supplies were met by the Three Mile camp foreman, Norm Waters, and his assistant, Allan Newell, in an old six-wheel-drive army truck. The presence of the Scot and the New Zealander was not questioned; they were just two more workers being added to the effort. A fierce wind was lashing the landscape with rain and sleet and the newcomers grimaced when they were directed to climb onto the exposed truck tray. Their destination, Three Mile Dam, was the staging point for roadworks and for the construction of Tumut Pond camp, sited several kilometres away at the bottom of a deep valley where Clear Creek ran into the Tumut River. Further downstream from this point Tumut Pond Dam would be built, and below that—370 metres deep inside the mountain itself—a ten-storey-high underground power station was to be constructed within a vast cavern blasted and hollowed from the living rock.

The two men clung to the frame behind the cab, burying their chins into their chests to protect their faces from the driving rain. The truck stopped at the foot of a steep hill, the 'Roaring Mag' which led to the camp. Despite the truck's six-wheel drive the

CONVERGENCE

The road to Guthega, 1951.

This signpost on the desolate high plain out from Kiandra marked the junction of the track to Tumut Pond and several old stock routes. It was the work of one of the bulldozer drivers and the name remains even on modern maps.

road was too steep for it. The men were ordered off and told to walk the rest of the way. They arrived about half an hour later, soaked and frozen, just as it began to snow. Time pressures did not allow for newcomers to be eased gently into the work or the environment. Jock and his companion were brusquely sent to a hut to collect straw for palliasses and then they were directed to their accommodation, which consisted of thin canvas tents. There were two men and two beds to each tent and, despite the raging blizzard, there was no heating.

Jock went to his bed cold, wet, hungry and extremely miserable. Someone somewhere had told him that Australia was hot. Yet here he was in a snowstorm, on a straw mattress in a tent with no floor, slowly freezing to death. He had never felt more wretched in his life and wished that he had never got off the train; or at least that he had had the sense to accept he was a new Australian and gone with the rest to Jindabyne, which he was convinced must be better than this.

The next morning he was assigned to a work gang which was to trek back down the Roaring Mag to widen the track. They worked without a break until five o'clock when they trudged back up the steep, muddy hill to what Jock hoped would be a hot meal, hot shower and warm, dry bed. But further delights awaited.

The ablution block was a concrete pad on which was erected a simple hessian screen nailed to wooden posts. The showers were canvas water bags suspended from overhead beams. The water was icy. Jock survived the chilling exposure and at nine o'clock queued for his first meal in almost two days. Mollified by the hot food in his stomach, he went to his tent where, by the light of a kerosene lamp, he discovered that during the day flies had laid maggots all over his damp blankets. He suddenly yearned for Scotland and home, so far, far away.

Jock was at Three Mile camp for three months before being transferred to another staging camp, set up at Adaminaby for the construction of the Eucumbene Dam. Like everyone else, he worked six days a week and was kept busy on the seventh with cleaning, washing and letter writing. Occasionally he would wander across to the other side of Three Mile Dam to yarn with an old prospector who was still eking out an existence from a mineshaft.

While at Three Mile camp, Jock worked as an offsider on an old D8 bulldozer (bought by the Authority at a disposal sale) until a contractor, Bob Dreise, introduced a newer HD 19 Allis Chalmers. It was Bob Dreise who, on arriving at the top of the hill near Three Mile camp where the access track and old stock routes met and branched in several directions, exclaimed in mock surprise, 'What's this—Kings Cross?' The name stuck and still appears on current maps.

The harsh conditions and the need, in the first twelve months, to rely mainly on Australian workers contributed to a rapid workforce turnover rate. By the time he had been several months on the Scheme, Jock was an old hand and, for the benefit of awestruck newcomers, was calculatingly blase about the hazards, particularly on the mountain roads. As a result of time constraints, the roads were unsealed access tracks, built just wide enough to accommodate trucks. Steep and cut into the sides of sheer slopes, they were extremely perilous. In winter they iced over and during the spring thaw turned into deep rivers of mud.

The road to the Tumut Pond dam site, May 1957.

One day in February 1950 Jock was despatched from Adaminaby to take stores, gelignite and a group of new arrivals, first to Three Mile, then down the steep, twisting road cut into the mountainside to Tumut Pond campsite. The loose surface and sheer drop for hundreds of metres made it a nerve-racking ride even under the best conditions. On this trip, Jock had already tested the faith of his passengers at the start of the descent by having continuously to fight the gear stick to stop it jumping into a higher ratio. Holding the truck in low gear with one hand and steering with the other, Jock crawled it along the increasing angle of the slope, touching the brake every few metres to check the speed. It was all very routine until Jock's foot suddenly went all the way to the floor as the brakes failed. The truck immediately gathered speed.

Jock negotiated the first bend without too much trouble but by the next hairpin was forced to abandon his hold on the gearbox and use both hands to wrestle the runaway around the turn. He knew of an approaching flat stretch where he could use the handbrake to stop the vehicle, but his terrified passengers, on the back among crates of explosives, were unaware of such a contingency plan and believed their last moments to be imminent. They were yelling and thumping on the cab roof. Close to the wheels of the truck tree trunks, rising from ground far below the lip of the road, passed in a blur of bush colours. The men searched with terrified glances for safe places to jump but, at the speed the truck was travelling, they could well have hurled themselves into space.

After some extremely anxious moments Jock did pull up on the short flat section. There, under the disbelieving gaze of his passengers, who had assumed it was the end of the journey with that truck, Jock secured a length of chain around a roadside log and joined it to the tow bar. The watching men looked on with undisguised horror as Jock cheerfully reassured them that all would be well. His bright, thick Scottish accent sounded to them like the cackling of a maniac and his mischievous grin only confirmed them in their opinion. Only a deranged man would continue such a journey using a fallen tree as a brake.

With the log secured, Jock restarted the engine and nosed the truck off the flat and once more into its steep, giddy descent. The big log running lengthways behind the truck bit immediately into the road surface, checking the vehicle's speed but not enough to reassure the men sitting nervously on the back. Jock watched gleefully in his rear-vision mirror as one by one his passengers grabbed their packs and leapt from the jolting truck tray until he alone was left to finish the ride.

The mountain roads claimed many lives, forcing the Authority in 1960 to fit seatbelts to all its vehicles. The Snowy Mountains Hydro-Electric Authority is reputed to have been the first organisation in the world to introduce the compulsory wearing of seatbelts. The reduction in serious injuries and fatalities achieved as a result of this measure led the state governments a few years later to extend the law to all road users across the country.

The introduction of the seatbelts was an example of Hudson's attention to details that others missed. The seatbelts were ordered after an accident in Norway in which Snowy engineer, Ken Andrews, was involved. Andrews and a Norwegian engineer were in a Volvo which was involved in a head-on collision with another car. The pair were wearing seatbelts, which had become a standard fitting in new Volvos, and were uninjured; the unrestrained driver and passengers of the other car were seriously hurt. On his return

CONVERGENCE

Three Mile Camp, 1950.

Andrews made only a passing reference to the incident but Hudson immediately grasped its potential significance and sought a full account from the engineer. Hudson was desperate to reduce the bloody toll on the Authority's roads. After listening to Andrews recount the episode in detail, he dictated there and then a memo authorising the importation of seatbelts for all Authority vehicles. Wearing them was made compulsory, and dismissal was the penalty for failure to comply.

As Jock Wilson was settling into life at Three Mile camp, Major Clews and his surveyors were camped in the valley below, surveying the site for Tumut Pond Dam. The survey team had spread itself across both banks of Clear Creek, a tributary which entered the Tumut River just upstream from the flat. They had built a wooden footbridge to link the two small clusters of canvas and it was the major's practice to rise at dawn, cross the footbridge and rouse his charges for the day's labour.

One April morning in 1950 he had just stirred the opposite camp into life when a big log being swept down the fast-flowing creek smashed into the bridge, carrying away its broken timbers in the turbulent white water. The creek was only a few metres wide, but ten days of rain had turned it into a torrent which was far too deep and fast-flowing to wade. Fortunately several men, including the guide and packhorse handler, Max Paterick, were already on the bank of the main camp. To recross the creek the men made a ferry by rigging a pulley and rope to an inflatable rubber dinghy they used for river

Major Clews overseeing survey plotting at Clear Creek, April 1950.

crossings. The crossing was wild and wet but Clews and Paterick supervised the transfer of all men and their gear, until only the major remained to cross.

He climbed gingerly into the dinghy and waved for the men to start pulling. Just a couple of metres out from the bank the rope jammed in one of the pulley blocks. The dinghy, suddenly rigid in the torrent, flipped upside down and the men on the bank watched in horror as the major's 'mounties' hat floated rapidly downstream until it disappeared. Assuming the major and his cherished hat were still as one, several men scrambled through the scrub, trying to keep the hat in sight. Meanwhile, the others gloomily freed the offending rope and dragged the upturned dinghy ashore. As they hauled it into the bank a faint choking sound was audible beneath the craft. Quickly turning it over they found the major still there, secured to the dinghy's floor by a section of rope which had caught tightly across his neck. Though it had nearly strangled him, it had held his head out of the water.

The concerned workers helped the old man to his feet. He was loudly spitting water and glaring wildly about to see if he could detect a culprit. It was then he noticed he had lost his hat. He pursed his lips in a moment of thought, then barked for someone to get his rum. Duly comforted by a tot, he squared his shoulders, strode to his tent and issued instructions that he was staying there and that the camp was closed until he had a new 'mounties' hat.

The men, mostly new Australians, were at a loss what to do. No one was sure if he was serious or joking. But the major was the major and they had certainly never seen him so upset. So they radioed his demand to Cooma. In Cooma it was greeted with equal puzzlement and the message passed on to Hudson in Sydney. No one knows what transpired, except that four days later a jeep arrived with a new Canadian mounted policeman's hat. The major then emerged from his retreat, resplendent in his new headgear and smiling happily as though nothing untoward had ever occurred.

Close behind the surveyors and road builders were the first work gangs assembled to prepare construction sites. Their campsites were planted on ground hacked into the sides of the mountains—at Jindabyne, Munyang, Guthega, Smiggin Holes, Tumut Pond, Tolbar, Happy Jacks and numerous other sites. During the twenty-four years of construction, almost 100 000 men are estimated to have lived and worked in the mountains in more than 120 camps.

Initially most camps comprised army-surplus tents with no flooring and furnished only with wire bunks on which the men put straw palliasses. The more enterprising workers made tar paper with which to line their tents against the cold. Later, Italian carpenters were brought in to build wooden barracks and erect corrugated iron Nissen huts. Each worker had to provide his own blankets, eating utensils and crockery. On-site contractors provided meals for about two pounds (twenty per cent of the base weekly wage) a week.

The conditions were primitive and in camps established in the gorges, such as those along the Tumut River, the funnel effect could drive winds up to speeds of 140 kilometres an hour. Such winds could pick up full fuel drums and hurtle them through the air like deadly missiles.

20 YEARS – 121 CAMPS AND WORK SITES

COOMA
1. Igloo-Tent camp (Mittagang Road)
2. West Camp
3. North Cooma
4. East Cooma

SNOWY RIVER SHIRE
5. Bolairo
6. Wambrook
7. Wambrook Radio
8. Tin Mine

JINDABYNE
9. Mill Camp
10. Balt Camp
11. Jindabyne Dam

KOSCIUSKO
12. Island Bend Road Camp
13. Island Bend Bottom Camp
14. Island Bend River Camp
15. Island Bend
16. Crackenback
17. Smiggin Holes
18. Guthega
19. Spencers Creek
20. Spencers Creek Survey
21. Munyang
22. Munyang Adit
23. Pipers Creek
24. Perisher
25. Nimmo
26. White's River
27. Gungarlin
28. Valentines
29. Dicky Cooper
30. Diggers Creek

TUMUT
31. Brindabella
32. Yarrangobilly Village
33. Peppercorn
34. Lobs Hole
35. Clover Flat
36. Yellow Bog
37. Paddy's River
38. Blowering Township
39. Pound Creek
40. T2 Drillers
41. Kiandra 3 mile (tents)
42. Kiandra 3 mile new site (barracks)
43. Saddle Camp
44. Ginninderra
45. Little Plain
46. Dry Dam
47. Barney's Creek
48. Kiandra Chalet (snow clear)
49. Tumut Pond (survey)
50. Tumut Pond (Clear Creek)
51. Cabramurra
52. Mudholes
53. Ravine
54. Coultons Camp
55. Cumberland
56. Section Creek
57. Sue City
58. Kings Cross
59. O'Hares
60. River Camp
61. Kenny's Knob
62. Happy Valley
63. T1 Adit
64. Ogilvies Creek
65. Thiess Village
66. Deep Creek
67. Burns Creek
68. Alpine Creek
69. Gang Gang
70. Nungar
71. Tantangara
72. Goodradigbee
73. Rules Point
74. Sparkes Creek
75. Pilot (Cobberas)
76. Byatts Hut
77. Yarrangobilly Caves
78. Toolong Hut
79. Providence
80. Patons
81. Bradley's Hut
82. Teviot Bank

ADAMINABY
83. Adaminaby (old town)
84. New Adaminaby
85. Addicumbene
86. Bugtown
87. Staging Camp
88. Coles Camp
89. Adaminaby (electrical)
90. Old Dam Site
91. Tolbar Gate
92. Tolbar
93. Frogs Hollow
94. Eucumbene Portal (Township)
95. Junction Shaft
96. Happy Jacks (Village)
97. Bald Mountain Creek

GEEHI
98. Bundilla
99. Friday Flat
100. Tom Groggin
101. Indi
102. Old Geehi (Camp)
103. Geehi Works Centre
104. Seven Mile
105. Khancoban Back Creek
106. Reids Flat
107. Windy Creek
108. Grey Hills
109. Bradney's Gap
110. Khancoban Township
111. Leather Barrel
112. Bogong Creek
113. The Twins
114. Siberia
115. Broken Back
116. M1 Surge Tank
117. Bella Vista
118. Verandah Camp
119. Anderson's Spur
120. Middle Camp
121. Talbingo Township

CONVERGENCE

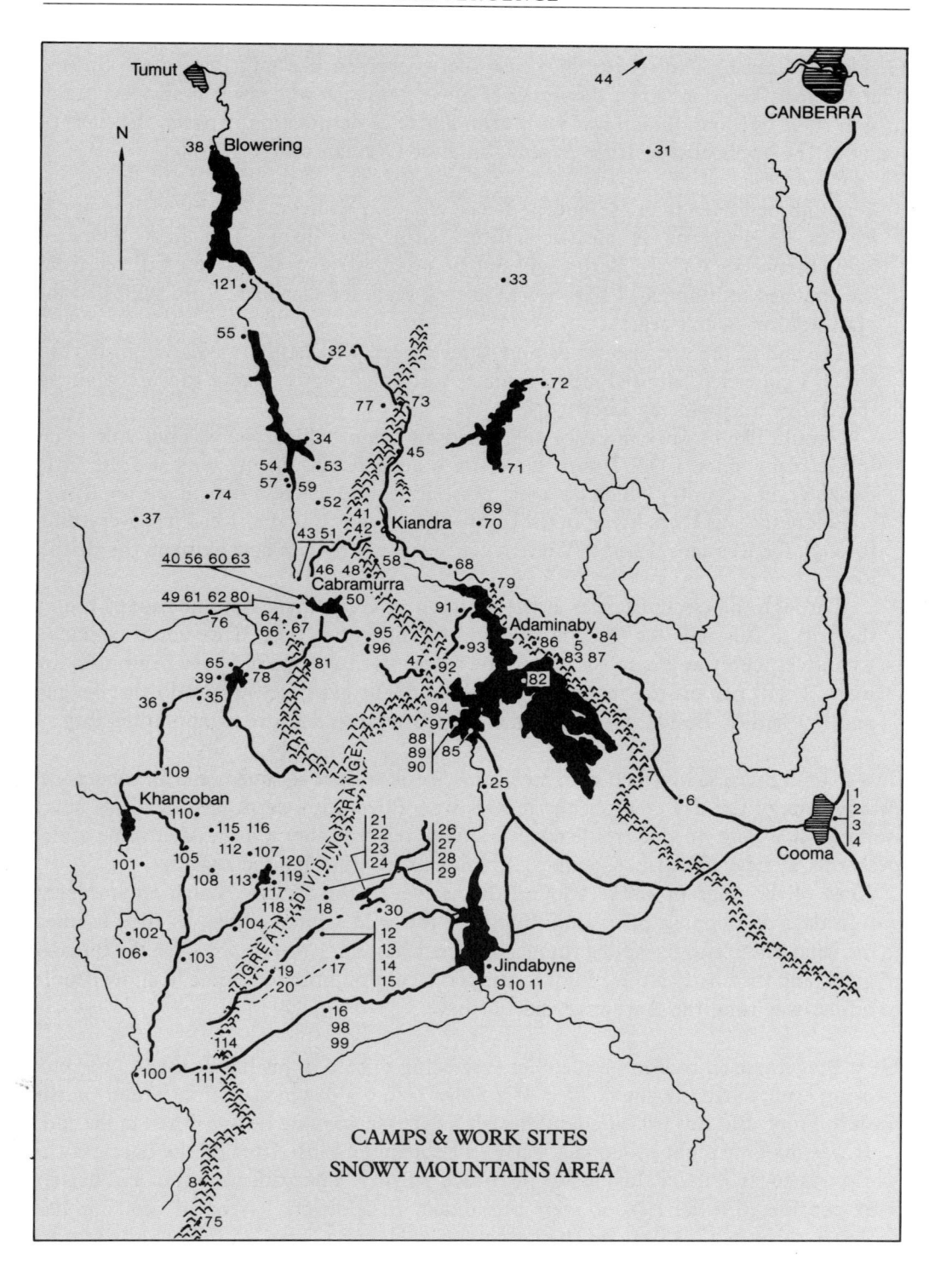

**CAMPS & WORK SITES
SNOWY MOUNTAINS AREA**

As the Scheme progressed and word of it spread, men from all over the world, prepared to ride the hardships in return for a new life, converged like pilgrims on the Snowy. The Authority began to receive thousands of letters from men who saw it as an opportunity to gain their freedom. There was more than a hint of desperation between the lines of many of the applications. Kurt Arnold, an East German refugee, wrote:

> Through pen friends in Melbourne I received your address. I hereby wish to apply for an employment as electrician/fitter with your firm. The firm Siemens-Schuckertwerke A.G. in Berlin will not be unknown to you. As you will see from the attached testimonial, I have worked seven years for that firm. I am specialist for high tension-switch gears...
>
> The end of the war also set an end to my career. The Military Mission in Berlin, which I contacted, advised me to contact Australia directly. I want to work again so that I can fully use my knowledge.
>
> I would like to work in peace and quietness again, which one was not able to do in Germany since 1933. I want to ensure a good future for my wife and my little daughter, in a country which has great possibilities and wherein free people are living.
>
> Before the war I was living in the centre of Germany but after I had lost everything through the Russians, I fled to Western Germany, where I am now living in the British Zone.
>
> Through the circumstances at the moment in Germany, I have no money to pay the trip to Australia for me, my wife and my little daughter. If we could conclude a contract, can you please advise me the money for the trip. If I have beginning my work, I will pay off the money to you. I would be glad, if you can do that for me and my family. Beside my testimonial I enclose an indenture of apprenticeship.

The spartan barracks to which such men came, alone at first, became home for thousands. At any time of the day or night the messes were filled with warm, smoky air as noisy workers sat at long wooden trestle tables eating or relaxing after a shift. After some initial problems at East Camp in Cooma, all nationalities were housed together.

Most of the migrant workers found themselves in this strange, harsh environment within days of stepping off a ship. Through the wild winter nights, as they listened to the wind biting and screaming through the branches, some men heard again the thunder of guns, and thrashed on their bunks, gripped by old nightmares. Time to allow people to adjust was time the Authority did not have.

Hein Bergerhausen couldn't remember ever being so cold in his life. It used to be cold, freezing even, sitting in the cockpit of a Stuka before a dawn take-off in winter on the Eastern Front. But this felt different; this felt lethal. He was sure he was frozen to the core.

It was his first night in Cooma, early in September 1951. Outside the barracks the world was encased in ice. Inside was not much warmer. The walls were thin and heaters were considered a fire risk, so were prohibited. In addition, it seemed someone had forgotten to supply blankets for the new arrivals. Hein had crawled fully clothed under the mattress and lay huddled on the wire base of his bunk, where he gritted his teeth and wondered just what he had got himself into. It was nightmarish and, judging by the noisy shivering and swearing elsewhere in the darkened barn of a room, he was not alone in his torment.

The following morning the camp foreman, an Australian, cheerily asked the newcomers how they had slept. He then told them he could organise blankets—but for a price. To replace the worthless German currency most of them were carrying, the men had each been given two pounds when they arrived at Melbourne. They gladly passed it over in return for blankets. Some, who had already spent their two pounds on clothing and other essentials, arranged to have the money taken from their first pay packet. Later, when a few made their first excursions into the town, they discovered they had been charged double the price being asked for blankets in the local shops. The Germans knew a swindle when they saw it and decided among themselves to get even at the first opportunity.

After buying their bedding, the newcomers were marched to the Snowy employment office where they were assigned work. In contrast to the night, the day was bathed in brilliant sunshine. Many of the men, Hein included, gazed upwards in rapture. It was the most beautiful, cloudless blue they had ever seen.

They had also stared in disbelief at the horse rails, some with horses attached, fixed into the ground outside the hotels and some of the stores. It was a paradox they did not understand. Australia was clearly a rich country but by European standards it still seemed materially backward. They could not see why anybody in 1951 would still be riding a horse instead of driving a car.

Despite the bright, sunny days, the nights remained bitter, and the first item Hein bought with his first pay packet was a sleeping bag. Other workers began to utilise their skills to make the barracks more comfortable. Within a few weeks the German barracks were adorned with beautifully crafted furniture. In a short time almost every bunk had beside it chests of drawers, dressing tables, shelves, even wall units. Many came to admire, including senior Authority staff who were pleased by the enthusiasm the Germans were showing in settling into their new circumstances. No one thought to wonder just where the wood was coming from until one of the Italian building contractors started becoming obstreperous about the timber that was disappearing from his yard.

One morning the camp foreman stormed into the barracks: 'Right-oh you buggers, I know what you're up to.' He launched into a tirade, warning of the dire consequences for all when the commissioner, Mr Hudson, learned of the thefts. The Germans countered by suggesting the commissioner might also be interested to learn about the foreman's blanket deals. The Australian reluctantly agreed they had a point, but explained that there was still the problem of an aggrieved contractor who had to be placated. After some discussion, the Germans agreed to pay for the timber.

The next morning the foreman and the Italian contractor arrived. Together they moved from bunk to bunk calculating the value of wood in each man's pieces of furniture. The Germans paid up, but at prices which they believed to be rather exorbitant. 'But we couldn't really complain. We did, after all, pinch the wood,' Hein recounted.

Hein was signed on as a diesel fitter at the Polo Flat workshops on Cooma's north-eastern outskirts. After lunch on his third day the leading hand, an Englishman, and his offsider, an Irishman, beckoned Hein to join them in the cab of a small truck. On the back was a large bulldozer radiator. Hein grabbed a box of tools and they drove about twenty kilometres out of town to a place called Rhine Falls, on the Adaminaby road. There they turned off onto a rough track which wound its way through thick scrub to a small clearing where a bulldozer was parked beneath a tree. Hein deduced, without any words needing to be spoken, that they were to change the machine's radiator. But as soon as he had unloaded his tools the Irishman told him, with a blend of English, sign language and a smattering of German, that he and the leading hand were going to Adaminaby. He tapped his wristwatch to indicate they would be back at five o'clock. Before Hein could protest that it was at least a two-man job, the pair drove cheerfully away.

When the rattle of the truck had faded, the silent bush wrapped itself around the nervous German. This was his first encounter with the Australian wilds and he had no idea what perils and beasts lurked within its inhospitable, dry-looking foliage. It was a bushland very different from the soft green forests of Germany.

He dragged his eyes away from the menacing scrub and back to the task at hand. Fortunately the dozer was under a sturdy limb, so with rope from its toolbox he was able to rig up a pulley system with which to lift and replace the heavy radiator. He worked busily and finished the job with time to spare. Still nervous about the surrounding bush, he planted his toolbox against the solid, comforting wall formed by the massive steel blade and sat down to await the return of the leading hand and his Irish mate.

Five o'clock passed. So did five thirty. By six o'clock Hein was sitting on his toolbox in the dark, a lone human in an alien world of strange, frightening sounds. He thought about walking back to Cooma, but realised that he did not know the way. He considered lighting a fire but, even though he had been in the country only a week, he was already conscious of the fear Australians had of bushfires. He also worried that a fire might attract bush animals, and he had no desire to confront any such creatures. So he curled up on the bulldozer seat and tried to sleep. It was hopeless to expect that he would sleep but he later confided to friends that he was 'too bloody frightened' to feel the cold.

Towards morning he did doze fitfully, only to be awakened at sun-up by a terrifying, ear-splitting screeching that widened his eyes and sent his heart thumping. Later, he learned to recognise the sounds as the sharp whiplash cries of currawongs and nesting magpies. Shortly after sun-up he heard the occasional sounds of distant vehicles. Reluctant to spend a minute longer than necessary where he was, he strode into the bush, trying his best to follow the track. Forty minutes later he emerged on a main road where he hitched a lift to Cooma.

Back at the workshops, Hein discovered that the Englishman and Irishman had spent the previous afternoon in the Adaminaby hotel, rollicking on into the night. They had forgotten all about the German they had dropped off in the bush. That, at least, was their story. The Australians present just laughed and slapped Hein on the back, congratulating the former Luftwaffe pilot for handling his first night in the bush with such stoicism.

East Camp, where the first migrant workers at Cooma were stationed, was a bare windswept hill just outside the main townsite. The camp comprised rows of wooden barracks, Nissen huts, a mess and an ablution block which, even in winter, was served

'Wages' accommodation at East Camp, Cooma, 1950.

with only cold water.

Apart from his night in the bush, Hein's first taste, literally, of Australian living was his first meal in the mess: 'There was so much meat. It was something everybody talked about! I'm not a big meat eater but I thought, "Oh well". So I held out my plate for some lamb. When I got back to my table I found it was still joined by a big tuft of wool. I lived on desserts for weeks afterwards and have not eaten lamb since.'

The Italians, however, had the hardest time with the English-style cooking. Spaghetti, spaghetti, spaghetti, they demanded. The Australian cooks had no idea what they were talking about. Finally, the Italian contracting firm engaged to build houses for the Authority staff set up its own mess, and had spaghetti sent from Melbourne.

We got on well with the Italians. Their mess was opposite our barracks. They were a rowdy bunch. My God, in the night time you thought they were killing each other. They were celebrating every bloody night. The noise was unbelievable. The local Australians were overwhelmed. They didn't know which way to run.

Occasionally Hein was despatched to work camps being set up in the mountains. Places like Island Bend and Perisher, primitive settlements of tents and wooden barracks, populated by hundreds of hard-working, hard-drinking men of a dozen nationalities.

When he arrived, Hein hadn't known what to expect, either at Cooma or on his sojourns to the mountain camps. After his experiences with the cold nights, it was a shock to discover men were living in the mountains in tents. As the population of workers in the camps grew, it seemed to Hein that he was surrounded by every nationality on Earth.

He devoted his spare time to learning English. Twice a week he went to night courses provided by the Authority; he took correspondence lessons and tuned into an English language program broadcast daily by the Australian Broadcasting Commission. To practise reading he took home lists of spare parts and machinery instruction booklets. After seven months his English was good enough to enable him to answer the phone, and he was appointed a leading hand. At the end of his first year he was promoted to foreman.

The first two years passed quickly and the time came for Hein to decide whether he would stay in Australia or return to Germany. He had grown to love the country. He liked Australians and their casual manner and was earning a healthy twelve pounds and eight shillings a week. He had been sending most of this to Cologne where life was still very hard. Australia by comparison was a paradise, yet he worried whether Sybille would like it. Cooma could be rough, and she would be a foreigner.

Despite her husband's reservations, Sybille was enthusiastic about coming to Australia, and Hein took his first holiday in two years when he travelled to the Melbourne docks to collect his wife and young daughter, Brigette. It was a happy reunion. Brigette was dressed in traditional costume. Hein was proud, but also a little concerned by the stares she received as they walked through Melbourne's streets. However, his anxiety was quelled the moment Sybille alighted from the train at Cooma and looked down over the valley and the town. She thought it was beautiful and felt instantly at home.

One of her strongest early impressions was the amount of food in the shops: 'There was so much, though it was very different to Germany. Australians seemed to exist

English classes in Cooma, 1951.

primarily on lamb or beef and vegetables and tea. There were no herbs, spices, salami, pressed meats or coffee. If you wanted coffee, you had to order it. But we were happy to live without it.'

Whereas many of the newcomers found the local Australians hostile and resentful, Sybille found them to be the most tolerant and helpful people she had ever met:

> I didn't know what a pound or shilling was. I had to open my purse and let them take out the right amount. As I learned the currency I became aware that no one ever took a penny more than the correct amount. When you walked along the street before the shops were open, goods which had been delivered by the freight truck or from the train were stacked outside. You couldn't have done that in Europe. It would have been stolen.

Before leaving Germany Hein had been warned that, as a German, he was likely to encounter bitterness. But he found only friendship:

> For the first few days I was worried... but you could see almost straight away there was nothing to worry about.
>
> Everyone just seemed to be glad to be here. Czech and Polish workers helped the

Germans learn English. At nights we played cards and shared our experiences. Everyone had plenty to talk about—how we had come to Australia, our plans to start a new life and how much we were looking forward to being reunited with our families. The separation was hard for everybody and the main goal was to work hard to be in a position to bring out our families as soon as possible.

It was a happy time though. The war was behind us; we were all starting again. The only time I was reminded of the war was on the first Anzac Day in the mountains in 1952. A Dutch engineer took a few Germans aside and explained what Anzac Day was and suggested we should stay away: 'It is the Australian soldiers' big day and you could get a punch on the nose,' he warned. So he sent us into Cooma to do some maintenance on the old diesel power station.

While the different nationalities got on together, it was no kindergarten. It was a rough place and there were a lot of fights. But they were not because of a person's nationality—except when the Serbs and Croats went for each other. They were the only ones who could never get on. We never were able to understand why.

The only experiences the Authority itself had of racial strife was between Poles and Germans: one was in the first few weeks after the arrival of a large contingent of Poles at East Camp; and the other at a road construction camp called Wambrook, just out from Cooma on Adaminaby road.

The Poles, from the moment they learned of the Germans' presence, gathered nightly in a large group, armed themselves with bricks, lumps of wood and even shotguns and marched on the German barracks. Every night, the residents of the once sleepy little town were awakened to the sounds of a riot as about two hundred Poles launched bricks and rocks into the German quarters, yelled obscenities and cries of vengeance and fired shotguns into the air. In the still of the night, the noise reverberated through the valley. Every night the police, led by a recently arrived young detective, Bill Holmes, dashed to the camp and, after some boisterous arguing and shoving, managed to settle everybody down and herd the Poles back to their own barracks. In the end Bill Holmes telephoned the commissioner, Bill Hudson, and pleaded with him to abandon his initial decision to segregate the nationalities. Hudson obstinately refused, arguing that it was better to keep language groups together.

It took three weeks, a mounting pile of broken glass, and increasing complaints from the town to force Hudson to relent and issue the order for all nationalities to be randomly mixed. There was no more trouble over nationality from that moment on.

At Wambrook, the German engineer Walter Hartwig, who had helped recruit many of the tradesmen, was called in to settle a conflict between Germans and Poles. He called the two groups together and spoke simply and to the point. 'We are all new Australians,' he told them. 'This is an honorable title which is not to be abused. The nonsense of Europe has no place here.'

One of the first major construction camps established in the mountains was at Island Bend, built as a base for the Guthega dam, tunnel, pressure pipeline and power station.

Water impounded by the dam on the upper Snowy River, where it met the Guthega

River, was to be delivered through a five-kilometre pressure tunnel to a surge tank, at which point it would enter a pressure pipeline. The water would plunge almost a kilometre down this pipeline to the power station's turbines.

The twelve-million-dollar contract for the Guthega project—the first such contract of the Scheme—was awarded to a Norwegian firm, Selmer Engineering, in September 1951. But before the Norwegians could put their expertise in hydro-electricity into effect, the construction sites had to be prepared. Most of the workers and tradesmen sent there were migrants.

Guthega, Munyang and Island Bend—names totally unknown to Australians at large—grew rapidly into bustling multinational communities. They were separated by just a few kilometres. Guthega camp was built for the construction of the dam; Munyang camp was for the power station; and Island Bend became the main Authority town for the Guthega project, and later the Island Bend dam and tunnels. At first Germans accounted for most of the population; then, when Selmer Engineering moved in, Norwegians arrived; and finally, as Hein Bergerhausen noted, there was 'the whole darn lot'.

The camps were remote outposts of civilisation, populated by hard men who set their own laws. A man who worked hard and could handle himself with his fists was a man respected, regardless of nationality.

It was a community of strange circumstances and associations. The generating plant that supplied power was driven by two diesel motors from a U-boat. There was no doctor, but a veterinarian from Jindabyne became a trusted healer of minor ailments. And the settlements, with their large population of lonely men earning high wages, became the first focus of activity for prostitutes from Sydney's Kings Cross.

Island Bend barracks, 1951. Primitive but an improvement on living in tents.

Business for the girls was good—in fact too good—and the Sydney vice squad, or 'flying squad' as it was known, was instructed to stamp out their activities in the mountains. The Scheme was supposed to be a showpiece of Australian industry and ingenuity, but not of the nefarious kind.

The flying squad hatched a simple plan. They would follow the girls from Sydney and raid Guthega camp with sufficient force to deter both the girls and the workers from carrying on the illicit trade. Squeezed into a shiny new Holden and brimming with confidence, four detectives tailed several carloads of girls from Sydney on the eve of a pay weekend. They followed them to Jindabyne, then backed off to allow time for business to begin. After a suitable wait, the policemen drove up the winding road through the Perisher Valley. At the turn-off to Island Bend they passed a gatehouse, from which a watchman waved a nonchalant greeting.

Unknown to the policemen, the watchman immediately phoned the camp to warn of the impending arrival. The workers, prepared for such an eventuality, despatched the girls with bundles of blankets to lie low in the surrounding snow-covered bush, while they went down to the construction site. When the detectives drove into the camp it was deserted. The officers began working their way through the barracks, pushing open door after door to find empty rooms. They soon realised they had been outmanoeuvred. Angered by this unexpected turn of events, they grabbed a man who had been following their searches and demanded to know the girls' whereabouts.

The man played dumb. Confident of their authority, the detectives dragged the fellow into a shower block, where they proceeded to interrogate him, using the techniques known to plain clothes policemen the world over. In this case, however, it was a serious tactical error. Others had seen the struggling captive dragged into the showers and sent word to the construction site for help.

Within minutes, a group of furious workers—among them battle-hardened ex-Waffen SS soldiers—descended on the hapless detectives. First they dealt with the nice new Holden, puncturing its tyres and pushing it over an embankment. They then stormed into the shower block, emerging a short time later with four bloodied and bruised detectives, whom they tossed unceremoniously into a snow drift. The officers staggered from the camp in a state of shock and began the long walk back to Jindabyne.

According to an unwritten camp law, a gang attack on a single man was a cowardly and intolerable act. Under the frontier justice of the isolated workcamps, it was irrelevant that the assailants were policemen exercising the authority of the state.

After the encounter between the Guthega workers and police, talks were held between the Authority and the New South Wales Police Department. These resulted in a private agreement that prostitutes could visit, subject to certain conditions set down by the local police. As prostitution was illegal neither the Authority, nor the police, nor the politicians could condone it publicly.

About the same time, the Authority came to an arrangement with the federal police to set up a unit of patrol officers with the status of 'special Commonwealth constables', to police the camps. This proved an effective system which enabled the Authority to keep the lid on petty crime, mainly through the threat of instant dismissal. The migrant workers in particular were more concerned about losing their jobs than spending a few nights in a police cell.

The rules governing the activities of prostitutes were eventually laid down by Cooma detective, Bill Holmes, who decreed that the women were allowed in the camps only after dark on Fridays and Saturdays, and only on pay weekends. To his critics, whose sense of morality had been affronted, Holmes pointed out that there was no choice. 'It wasn't a Sunday school being run up in the mountains; and it was one means of keeping the men in the camps,' he said. 'We only had six police in Cooma and the place was already enough of a madhouse, without suddenly having an extra 4000 men, many of whom were pretty rough characters, descending on you every weekend.'

The men themselves, who knew that any breach of the rules would only cause unwanted strife, were given the chance to regulate the prostitution. The organisation of the girls created some interesting sights and situations. A particular hut or barrack would be requisitioned and the girl, or girls, accommodated. Then the men would line up outside, with sometimes twenty or thirty men queuing to visit a single girl.

Because the girls had to be gone by daylight, there was no time for small talk or extended intimacies; but even a few minutes with a prostitute broke the hearts of some men, lonely and far from home. The Italians were notorious for falling in love, then demanding fidelity on pain of death. Time and again some aggrieved Romeo had to be consoled or subdued by his workmates before he ruined the arrangements for everybody.

By and large, however, it was an orderly business. The girls clearly made a lot of money, and after a while many began to fly down from Sydney for a weekend's work. Occasionally the rules were stretched to accommodate the needs of night-shift workers who could only see the girls during the day. But when word spread that the girls were staying longer than had been stipulated, there would be a raid by the flying squad. These daylight raids provided great entertainment for the rest of the camp as constables and detectives tried to dislodge reluctant 'ladies of the night' from hiding places beneath the barracks.

The presence of ex-Waffen SS in the camp was common knowledge at the time, though there are some people who still deny their presence on the Scheme. The SS practice of tattooing men's bloodgroups under their armpits made these people clearly identifiable in the communal showers of the Snowy camps. Once they had been cleared by the immigration screening procedures, few of them felt any need to hide their identities, particularly as they were living in such remote locations. Nevertheless, some of them had had skin grafts in an attempt to hide their tattoos.

Among the former Waffen SS at Guthega was 23-year-old Karl Maier, who was trying desperately to forget the past and rebuild a shattered life. Coming in towards the war's close, Karl was probably not the typical SS trooper; but his experience demonstrated that the enemy was not always what he appeared.

In late 1944, when the war had turned against Germany, the Reich government was becoming desperate for mechanics and engineers to keep its war machines moving. At just fifteen years of age, Karl was plucked from his studies as a trainee diesel mechanic at a technical college and assigned to a Tiger tank in a Waffen SS unit.

The passing of forty-five years has not dimmed the searing memories:

I was a boy, a student with no rank, no uniform even, just a sergeant with a gun to make me do as I was told. It was hell—a nightmare that I have never told anyone before. It is not something that is easy to put into words.

These men were fanatics and prepared to die fighting for the Führer. Towards the end we were just rampaging, travelling by night and driving our tanks into houses during the day for cover. Sometimes we had skirmishes with Americans but more often we were shooting German soldiers who were retreating. It was something I tried very hard to forget, but it is difficult when your body bears the scars of bullets put there when you were just a boy.

The end came when we were surrounded by an American unit. Our officer wanted to go into the mountains and fight on, fight and kill until the last one of us was dead. But the rest had finally had enough, so someone shot him. When we surrendered, and the Americans saw the unit they had captured, they just lined everybody up and shot them on the spot. No questions, no trial, just revenge.

I was spared because I was a boy. They asked me what I was doing with that unit. I told them I'd had no choice, that I'd been pulled from school and put in a tank. It was still enough to earn me six months hard labour in a French prison and brand me for life.

After serving the sentence Karl returned to Germany, bitter and full of hate. Everywhere he looked he kept seeing the same arrogance that had fuelled the war in the first place. He wanted to see people feeling as he had felt, suffering as he had suffered; to see German pride at least subdued by the bruising it had taken. Instead it was build, build, build; go, go, go; dance to the tune of the American masters. People talked of peace. What peace? Certainly there was no peace of mind for young men like Karl.

Normality was slow in returning. In its aftermath the war still reached out to kill and maim. Mines, booby traps, shells, guns and handgrenades still lay where they had been abandoned or hidden in forgotten caches. In most towns and villages there was no civil law and order for almost a year. In that time the law of the jungle reigned. Black marketeers waged deadly entrepreneurial wars. Russians, Poles, Czechs and other East Europeans, freed from prisoner-of-war and concentration camps, wandered aimlessly. They had no homes to return to and were forced to steal, loot, and sometimes kill, to survive. The disorder was total, the confusion complete.

At home, Karl's father insisted that his son join him at the diesel motor factory where he worked as an engineer. 'You must settle down. You owe it to Germany. You have a responsibility to help Germany rebuild,' he told him. Karl always responded with a bitter laugh: 'I've been kicked, bashed, trodden on, shot at, wounded, been forced to kill and you think I owe Germany something. Well, I don't.'

Karl shunned all efforts to rein him in, deciding instead to take his chances in the black market. He became quite successful:

I'd get aboard trains as they slowed through towns and simply grab whatever boxes I could lift and carry off.

One day I took a box which I later found was full of new American Colt 45 sidearms.

Another time it was some officer's clothing, which I had made into a handsome suit. It was the smartest suit I had ever owned. But the first night I wore it on the town to meet my girlfriend the khaki material was recognised by Americans. They tore it from me, there in the street. I was left waiting for my girl in nothing but my underpants.

Such was life in a world still touched by the insanity it had created. But Karl was living only in the present. Such things had no bearing on his future because he did not believe there was a future—only a nightmare in his past, for which he still sought revenge from the society around him.

The black market was one way of getting back at the bastards but as the police and military became more organised that finally became too dangerous. I eventually decided to try to start again and resumed my mechanical training with MAN Diesel.

I was restless but they were understanding. They learned what had happened to me in the war and went out of their way to settle me down—encouraged me to find a wife and even allocated a new flat to me. It wasn't what I wanted but they were trying so hard that I felt I should at least make an effort. I went to have a look at the flat. They were scarce and I knew I should feel privileged. It had a brass plaque on the door with my name and title 'senior mechanic'. While I was looking at it a women from the next flat came and said: 'You look too young to be a 'senior' mechanic.' I snapped. It was the same arrogance and obsession with titles and badges and pride that had started the whole fucking war. I went to MAN and told them they could shove their flat.

That's when they decided to send me to Australia, to instruct Australian mechanics on servicing their new motors.

From the moment Karl arrived in Australia and encountered his first Australians he felt a weight lift from his heart. Australians were friendly, warm-hearted, and they hated sham and hypocrisy. They had made a national pastime of cutting down the imperious and high-nosed with a razor-edged humour. Karl loved them and from the first day knew he would never return to Germany.

After six months, he left his job with MAN and joined the rough-house world of the Snowy, where for the first time in his life he was not only happy but even began to believe in a future. The hard work, hard drinking, humour and mateship of the Snowy was a cure for the afflictions of the mind.

Karl and others like him passed through the immigration screen because their war records could be checked and they were assessed as politically safe. Karl's unwilling association with the SS had made him vehemently anti-Nazi.

In any case the Australian government's policy towards former Nazis was relaxed surprisingly quickly after the war. A 'Top Secret' Cabinet minute, dated 20 November 1946, reported that the demand for German scientific expertise had grown to such an extent that the Russians had breached the de-Nazification pact made with their former allies and had started offering employment to German scientists regardless of their political history. Britain also was understood to be quietly employing former Nazi scientists. Cabinet was told that the consensus among Australia's scientific and business community was that Australia should take a similar long-term view and make full use of the best German scientific and technical brains.

The government therefore decided to offer employment to German scientists, irrespective of past Nazi links but subject to their 'strict supervision' while in Australia. Those who wanted to stay were put on nine months' probation. If after that period both parties—the scientist and the Australian government—were happy with each other, the government would pay to bring the scientist's family to Australia. After five years' residency, normal naturalisation procedures would apply. To circumvent the sensitive issue of a scientist's political history, the government introduced a rather vague and subjective screening procedure to establish that a person was 'politically unobjectionable'.

By 1949 the government's former resolve to keep out anybody with past Nazi links had become even further diluted. In response to a query from the Victorian trade union movement, the Minister for Immigration, Arthur Calwell, responded that immigration checks on Germans were now aimed at keeping out anybody who was an 'active' Nazi.

The Australian Security and Intelligence Organisation is understood to have a list of about 200 suspected war criminals who lived in Australia at some stage after the war. While it is possible that some of these migrants were knowingly sponsored by the federal government, it is unlikely that anybody who had been involved in atrocities would have sought work on the Snowy Scheme. The presence of so many Germans and former slave labour camp inmates would have significantly increased the risk of detection.

For the handful of Australians at places like Island Bend, the presence of the Germans, the activities of prostitutes and incidents such as the flying squad raid were windows into a totally new world. As an eighteen-year-old not long out of school, Bob Leech had his worldly education broadened beyond his wildest imagination.

Bob had moved to Cooma with his family after his father, Professor Tom Leech, had been appointed to set up the Authority's Scientific Services Division. It was his first year out of school, and Bob was working at Island Bend as a cement analyst. He was one of the few men in the entire camp who spoke English as a first language. At one stage he was the only Australian out of more than two hundred workers there.

As tents gave way to huts, Bob shared a double-decker bunk with an Irish tunneller in a poky two-metre square hut. The Irishman, who had the top bunk, used to hang his sodden, putrid socks from the bunk to dry. The stench was so bad that Bob was unable to sleep for two weeks. He finally obtained alternative accommodation by bribing the camp foreman with three bottles of gin. The gin cost him a week's wages, but a desperate situation demanded desperate measures.

In common with most of the Australians, Bob's first winter in the camp was his first experience of snow. Spending two hours each morning digging away the snow before starting work was an unexpected novelty. The hut doors swung inwards to expose a wall of snow, three to four metres deep, which the occupants had to breach before they could get out. Eventually the men found it easier simply to dig tunnels, rather than trenches, from hut to hut and out into the open. The camp barber set up business in an entire room excavated out of the snow beside his hut, using the cut hair to carpet the floor.

Wiser by the second winter, the men laid sheets of corrugated iron between the roofs of huts to speed and ease the construction of snow tunnels and effectively take the camp underground. The main danger then was fumes from kerosene heaters; it became a question of whether one died of poisoning or from the cold.

Mingling with the migrant workers, Bob sensed an undercurrent of aggression. He witnessed a number of bloody brawls which, however, often served to relieve the tension that had built up. But many of the violent incidents had serious ramifications.

Sometimes a man's mind would suddenly snap and trigger off a bloody, indiscriminate rampage. There was no attempt in those days to understand, or even identify, war trauma, or 'shell shock' as it was then known as. It is only since the Vietnam War that the realities and effects of war trauma have been widely recognised.

In one incident, Bob Leech recalled:

. . . one fellow, who seemed as sane as anybody, got depressed one day and just went crazy with an axe. He attacked the bloke he shared with, who managed to dodge the blow and get out, and then just started chopping into his hut—smashed it to pieces. They called up Cooma and the next day he was taken away in a straightjacket. You never heard what happened to blokes like that.

Among the workers based at Cooma was a former U-boat captain, who ended his days in an asylum in Melbourne.

Bob also witnessed the beginnings of the horrific accident toll as the pace of construction began to escalate:

The accidents were often so simple. I remember a fellow who decided on the spur of the moment that rather than walk to his station in the tunnel, he would ride on the locomotive. In the dark he didn't see a water pipe which had broken and slipped from its bracket on the roof. It speared straight through his head. When you saw something like that, it was pretty sickening and hard to forget.

After three years, Bob left the Scheme to become an agronomist. But he was far from finished with either the Snowy, or the mountains.

Back in Cooma, a world away from Guthega and Island Bend, the Authority was rapidly transforming the once quiet bush town, giving it an international focus. It had set up a Scientific Services Division, which was becoming home for some of the sharpest and most advanced minds in the country. The division was housed in a collection of Nissen huts on Cooma Back Creek, under the administration of Bob's father, Professor Tom Leech.

One of the young engineers under Professor Leech was Jack Lawson, a graduate from the University of Western Australia who had just gained a Ph.D. in hydraulics from Aberdeen University, Scotland. Lawson had joined the Authority in July 1951, signing on at its London recruiting office after learning of the research possibilities on the Scheme.

While the Authority was still shaping itself into a functioning body, it had sent Lawson on a tour of universities and research establishments throughout Europe to bring himself, and the Authority, up to date with the latest research in hydraulics. He travelled for six months from Scandinavia to Italy and found growing interest in the Snowy Mountains proposal. The size and scale of the project was capturing the imagination of engineers and scientists everywhere.

Jack Lawson finally arrived in Cooma in 1952, where he designed a fluids laboratory. With a Scottish colleague, Les Webster, he also built a test weir on the Cooma Creek, which despite the 'No Swimming' signs immediately became the local swimming hole.

To Jack Lawson and the many other young professionals, Cooma was raw but exciting:

Sometimes you felt you weren't living in Australia at all. At the start of winter, 1953, Les Webster organised a soccer competition—a sport then scorned by most native-born Australians. The blokes built a pitch and were immediately able to field six international teams—Dutch, German, Scottish, Italian, Yugoslav, English—and a multinational team of all-comers.

As an ex-Australian Rules league player in Perth, I was recruited as goalkeeper for the Scots. It became a keen and well-organised competition.

Jack Lawson (bottom right) works on a model of the Munyang–Snowy confluence in preparation for the design and construction of the Guthega power station and switching yard. In the background are the first buildings of the scientific services division.

Lawson was set the task of building two working hydraulic models of the Scheme. One of these was for the purpose of analysing the flooding risks posed by the confluence of the Snowy and Munyang rivers at the Guthega power station. This entailed building a detailed model of the area, including a precise representation of the geography and a model of the power station and switching yard. By simulating flood conditions, they were able to determine, through a series of exhaustive tests, the design factors that needed to be built into the construction and siting of the power station, its switching yard and other structures.

Every aspect of the Scheme involved detailed preparations. Nothing was left to chance. For example, at one point on the Upper Tumut River the design engineers could not decide whether to build a bridge or construct a culvert under a road crossing. Access had to be assured but costs also were important. It was only by getting Lawson to construct a detailed model of the situation and test it under a variety of simulated river conditions that the final decision, a culvert, was made.

One of the men whom Jack Lawson worked with was a model-maker named Frank Gibbs, a Londoner who during World War II worked on a number of top-secret projects. He made the models of the two Ruhr dams, the Moehne and the Eder, in preparation for the two destructive air raids of the 617 'Dam Buster' squadron. Frank, and other members of his 'dirty tricks' team, also made poisonous roses, exploding crucifixes and rubber bridges for use by European resistance workers. He once spent the night gutting hundreds of dead rats and filling their stomachs with explosive. The rats were to be strewn near the entrance gates to a German factory in the hope that overseers would throw them into the boilers—with devastating results.

The Snowy Authority found Frank Gibbs working as a set builder for a London film studio and immediately offered him the chance once again to build model dams, this time for peaceful purposes.

As an engineer, Jack Lawson was a member of the Scheme's higher social order. Much to his chagrin, however, he was forced by the lack of housing in the first year to live for six months among the workers at East Camp where there was no hot water, and cold showers on winter mornings made life unpleasant.

From the Scheme's very beginning there was a strict demarcation between salaried staff and wage-earners. East Camp, on one side of the railway line, was for the workers and Cooma North, on the other side, was for professional staff.

This often caused resentment among workers. Many of them were highly skilled professionals who were forced to work as labourers because (except for German engineers) the Australian government and academic authorities simply did not recognise their qualifications. Among the ranks of workers there were found doctors, lawyers, teachers, musicians, artists, criminologists and ballet dancers. There were, as well, a Greek classics scholar who gave recitals at meal times and an English peer, educated at Eton, whose job on the Snowy was fumigating gumboots.

Even so, Jack Lawson admitted to being rather upset when he was put in East Camp:

It felt degrading; but among the engineers I was a junior, so I had to take it.

The social divisions among the professional staff were complex. On the outside, it would have looked like a simple 'them' and 'us'; or 'staff' and 'wages', as everyone was tagged. But there was tremendous rivalry among the professionals. The social significance attached to the type of house a person or couple was allotted verged on paranoia. There were six or seven different types of house built for the staff, ranging from the small E-type, built by Dutch contractors, to the Italian-built Pasotti, a large bungalow with veranda. The type of house you were allocated depended on your status and people took this very seriously.

If people didn't know the status of their colleagues, they checked out the type of house they had been put in. So I felt pretty aggrieved at initially not getting a house at all.

After six months Jack was accommodated on the other side of the tracks, which enabled his wife, Olive, who had been waiting in Sydney, to join him. Olive, a highly regarded botanist and former lecturer at the University of Aberdeen, found Cooma strange and lonely. It was a far cry, not just from Scotland, but also from the stimulating academic set she had left behind. She very quickly felt the need to find a job.

Unfortunately, the Authority had yet to demonstrate any notions of equal opportunity, and women employees were almost non-existent. However, in one of those quirks of fate that can have such a telling ripple effect, the Scientific Services Division had been given the task of establishing a soil conservation unit because of concerns about the Scheme's environmental impact. Consequently, it needed a botanist.

Olive's first job was to build up a collection of all the eucalypts in the region—a mammoth task for someone who by her own admission barely knew a eucalypt from a bar of soap. But Olive's more immediate contribution to the future of the Scheme came from an unlikely turn of events.

It had slowly dawned on the general public, especially the inland fishing fraternity, that the Snowy intended punching large east-to-west tunnels through the Great Divide. This meant fish would, for the first time, be able to travel from eastern into western catchments. The fishermen knew there were eels up and down the east coast but no significant numbers of eels in the Murray River, the habitat of Murray cod. Eels are very partial to young Murray cod.

The New South Wales Government seized on this revelation. The commissioner, Hudson, received a terse letter from the New South Wales Fisheries Department pointing out that these large tunnels would make it possible for eels to make their way from the eastern Snowy catchment to the Murray catchment. The department warned that this was not to happen at any cost and demanded that he 'do something about it'.

The commissioner felt there were more important matters needing his attention, so he offloaded the problem onto Professor Leech at the Scientific Services Division. Leech had no idea what to do. However, while mulling the matter over, he grabbed a dictionary and found that the word for the study of fish was 'ichthyology'. It gave him an idea—not about how to solve the problem, but about how to unload it onto someone else.

He started ringing around the various people under his command. The first person bright, or foolish, enough to know what ichthyology meant would have the 'pleasure' of solving the eel problem. When eventually he rang Olive Lawson, she promptly gave the correct answer, and got the job. It was quite a task and it was made clear that failure to solve the problem could create yet another political and environmental headache, something the Authority could well do without.

Olive, being a botanist and not an ichthyologist, had first to learn precisely what it was she was dealing with. She ordered in every book on eels she could trace and engaged eel catchers. Eels were found, looked at, read about and advice was sought from anyone who claimed some knowledge of their habits.

One of the first ideas put forward was to place a series of electrodes across the mouth of each tunnel to electrocute any eel that entered. Advice was sought from overseas eel experts who immediately scotched the idea, saying it would not be reliable as some would merely be stunned, not killed, by the shock. Also there was the problem of zapping fish, a consequence that was unlikely to please either fishermen or the Department of Fisheries.

Everyone on the Scheme, it seemed, had an idea, but it was a hydraulics engineer who one day asked how the eels would cope with passing through turbines. There would be a massive drop in pressure from the upside inflow to the downside discharge from the turbine. No one knew the answer. An experiment was set up using a large pipe, flanged on both ends so that it could be sealed and pressurised. The instantaneous pressure drop was achieved by opening a quick-release valve at the bottom end. Eels were caught and the experiment activated over and over again at varying pressure levels. Not a single eel survived any experiment; the sudden pressure drop blew out their gills.

The saga ended then with the discovery that there was no problem. Any eel entering a tunnel would not survive the passage through the power station turbines. The department was sceptical; but still, forty years later, no eel is known to have made its way into the western waterways.

The morning sun was barely above the horizon as Mario Pighin pedalled the thirteen kilometres from his family's small farm to the nearby village of Berteolo in the northeast of Italy. Had anyone else been about at that early hour he or she may have wondered why the young man was dressed in his best suit. It was very simple: Mario was on his way to get married. The unusual hour—six o'clock in the morning—was necessary because he and his bride-to-be wanted to make the most of the only day they would have together as newlyweds.

On 9 May 1952 Mario had received a phone call from the local municipal offices telling him that the Snowy Mountains Hydro-Electric Authority representatives in Italy had advised that he would be sailing in just five days time for Australia. Mario rode immediately over to Berteolo to break the news to his fiancee, Angelina. 'So if we are to be married, it will have to be the day after tomorrow,' he told her. That was on a Monday night. The couple had half expected such a turn of events and had forewarned their local priest. They were married shortly after first light on the following Wednesday.

After a simple ceremony, they and the best man caught a bus to the provincial capital, Udine, twenty kilometres away. There the three took lunch and enjoyed a movie. It was the only sort of honeymoon they could realistically expect. Early the next morning Mario left to spend the day farewelling friends and neighbours, while Angelina went out to begin her new life working on her husband's family's farm.

The following day Mario caught a train to Naples and boarded a ship to Australia, where he had been engaged to work as a painter in the fledgling Snowy township of Cabramurra—the highest settlement on the Australian continent. They were not sure when they would see each other again, but under the conditions of his contract it would be at least two years. The reason for the hurried wedding was the Authority's policy of not employing married men from overseas. Mario and Angelina decided to wait until his contract with the Authority was confirmed and then marry surreptitiously before he left.

Just a few kilometres away, in the same province, another young couple, not known to Mario and Angelina, had the same intention and were anxiously awaiting word from the Authority. For childhood sweethearts Guiditta Miane and Alfredo Fabbro, the call came just a fortnight later. They were married on 29 June, three days before Alfredo sailed, also for Cabramurra, where he had secured a job as a carpenter. The couple's plan was for Alfredo to take advantage of the high wages being paid and return with the savings in about five years. It was a tough way to start a marriage but reliable work was hard to find in postwar Italy.

After their wedding Guiditta went with Alfredo on a train to Udine. Then she rushed immediately back home because the train carrying Alfredo and other emigrant workers would pass back through the village on its way to Naples, allowing her to catch a final glimpse of her husband-of-one-day on his way to a distant land.

Just kilometres apart, but yet to meet, two young brides began the same ordeal of

The Cabramurra wet canteen 1954.

separation as their husbands sailed to the other side of the world to build their futures.

Standing almost two metres—or six foot four, as measurements then went—twenty-year-old Probationary Constable Bev Wales felt conspicuous and silly as he strode along the busy inner-city footpath in Sydney, scanning for jay-walkers. 'Pedestrian traffic duty' did not appeal to him. He was a country boy, born and bred on a farm near Yass in southern New South Wales. He felt nervous among so many people. He also felt ridiculous wearing the 'poofy' white helmet he had to don as part of the summer issue. But he knew there was worse to come. The sergeant had mentioned that his height would make him ideal for point duty. Considering the mayhem of city traffic, that seemed a job appropriate only for someone who had a death wish.

Thus, on this sunny October morning in 1952, Bev Wales was not happy. Tramping footpaths wasn't his idea of what the police force was about. He had joined so that he could uphold law and order—apprehend villains and keep the peace. As he crossed the road, a bobbing white buoy in a river of people, the buoy suddenly disappeared. Removing his helmet as though to mop his brow, the young constable had deftly tossed his headgear under an oncoming tram.

He returned immediately to central to report his unfortunate loss. The sergeant, wise in the ways of recalcitrant young constables, simply put Bev on a tram to the Redfern clothing depot to obtain another helmet. A few weeks later he was told to report on the following Monday morning to the sergeant in charge of traffic duties.

'I won't do it,' he told the sergeant, who ignored the remark. But Monday duly arrived and Probationary Constable Bev Wales did not report. 'I won't do traffic duties,' he said, when called again before the sergeant.

The police force was unaccustomed to probationary constables dictating what they would or would not do, and Bev was summarily ordered before Superintendent Johnston, a white-haired old fellow who was nicknamed 'Cottontop' by the lower ranks.

'Is it a fact you have been detailed to do traffic?' the superintendent asked.

'Yes sir.'

'Is it a fact you refused?'

'Yes sir.'

'Why?'

'I didn't come to the city from the country to join the police force to be rubbed up by taxis and trams, sir.'

Bev told the superintendent he would resign if he had to but would prefer to be sent back to the country. The superintendent's white eyebrows lifted: 'And where would you like to go—Narooma or Cooma?' he asked sarcastically. The question was a trap. Narooma, on the south coast, was every policeman's dream posting. To choose that would probably have been the end of Bev's short career with the force. Bev, however, wanted a hardship posting. 'Cooma would be fine sir,' he responded.

On 6 December 1952, while still a probationary constable, Bev Wales reported for duty at the Cooma police station. Taming lawless elements in the mountains was precisely the type of police work Bev aspired to. Within a few short years the man, with his sense of bush justice, his fists and his fearlessness had become a legend.

CHAPTER SIX

Guthega Dam, built to create the headwater for the Guthega power station, under construction, winter 1952.

TENTATIVE BEGINNINGS

TENTATIVE BEGINNINGS

Every day for six weeks Dieter Amelung, the East German refugee, watched as busload after busload of workers, noisy and jocular, left the Bonegilla migrants' camp near Albury to start jobs. Every day, week after week, it was always someone else being called up and given the exciting news about their work: where it would be and what they would be doing. He knew many were going to the mountains. After six weeks he began to wonder if there was work for him in Australia after all, because towards the end only about twelve men remained, and all were electricians.

Dieter recalled how the Australians had rejected his first application to migrate because they said Australia had no need of electricians. Yet most of the workers were being taken to a hydro-electricity scheme. Surely such a project needed electricians? Yet not once had anybody in authority mentioned the Scheme, either to him or the other electricians he was with. It was a puzzle, and the answer to it lay on the other side of the language barrier. There always remained so many information gaps whenever they were told anything. So Dieter filled in his time trying to improve his English.

The waiting ended after six weeks at Bonegilla. The men were assembled one morning and told they were going by train to Cooma.

It was November and the Monaro was dry and desolate-looking. Some of the men recalled learning at school that Australia was the world's driest continent and was mostly desert. Certainly the landscape stretching out beyond Cooma was not the type of terrain they would associate with hydro-electricity. The hill above the railway station was treeless and covered in a mat of dead grass. But what they did not know was that, beyond and hidden by the low hills on the opposite, western, side of the town—across the valley of red tin roofs—was the top of the Great Dividing Range.

At Cooma railway station the men were directed to the employment office where they learned, to their relief, that they had all been placed in jobs. When Dieter presented himself, the Australian in the office ran his finger down the sheet on his desk: 'Amelung. Yeah, you're going to Cabramurra. Right, back here for the bus at seven o'clock in the morning.' He repeated the order several times until Dieter, indicating he understood, asked, 'What is Cabramurra?' The man gave him a thin smile: 'Speak a bit of English do you? That's good. Cabramurra is up in the mountains—way out in the bush. You'll see.'

The Authority had chosen a location near a place called Dry Dam as the site of a town, Cabramurra, which would be the construction centre for the two Tumut underground power stations, as well as the Eucumbene-Tumut and Tooma-Tumut tunnels and Tantangara Dam. Dry Dam, like Three Mile Dam, was a small dam built by Chinese gold miners as a source of sluicing water in the days of the Kiandra goldrush. It was called Dry Dam because the miners breached its walls after the gold ran out.

Dieter, along with the others, was then directed to East Camp where he was allocated a bed for the night. On the walk up the hill Dieter wondered why he could see no pipes or power lines, or any of the usual signs of a hydro power station.

That night there seemed little to do, the town was quiet—in fact deserted—so he went along to an English class, where he was almost immediately distracted by a photographer who was busy with a battery of lights, cameras and tripods. Dieter was a keen photographer and watched the man with interest. He did not realise it at the time, but he had been captured on film. A year later he appeared in a special *National Geographic* magazine article on the Snowy Mountains Scheme.

The end of the journey, Dieter Amelung (seated foreground) and other workers from Europe on arrival at Cooma in 1953.

Dieter Amelung on the summit of Mount Kosciusko shortly after his arrival in Australia.

TENTATIVE BEGINNINGS

The following morning Dieter again found himself marvelling at the size of the country. The train journey from Melbourne to Albury had been the longest journey of his life—hour after hour of the same paddocks passing by. Now, as the old bus rattled its way westwards across the plains towards the mountains, it seemed much the same. But when three hours had passed and he had still not seen a power station he began to wonder what everyone was talking about. Because of the language void, he did not yet realise that the power stations were still to be built, and Cooma was the base for the Authority, not the site of the project.

It was five hours before he was deposited on the dusty street that ran through Cabramurra, which itself was still under construction. Dieter looked around in disbelief. He was clearly in the town centre but there was just one tin shed that seemed to be a general store, and a post office. To one side a few prefabricated houses were being assembled. The tiny settlement sat in the middle of a large dirt clearing. All the surrounding trees had been razed—so he later learned—as a precaution against bushfire. From where he stood he seemed to be looking out across the top of the mountains, across the roof of Australia. It was impossible to envisage it then, but the town would later have to be moved because in winter the hollow in which it was being built became a huge snow drift.

Dieter was hot and his skin prickled under dried sweat. A little fazed by his first sight of the town, his future home, he approached the first person he saw and asked in German for directions to the Snowy office. It was not until the man replied in German that Dieter realised his lapse, but he was grateful for being spared another struggle with unfamiliar words.

When he found the camp foreman he was bluntly told: 'No room here mate; you'll have to go to Kenny's Knob.' On a careering jeep ride down the steep, twisting bulldozer track that descended to Kenny's Knob, about three kilometres below Cabramurra, the foreman explained that a tunnel was being drilled deep down to the subterranean site that had been chosen for the excavation of the vast cavern that would house the Tumut-1 underground power station.

At the mention of the words 'power station' Dieter's spirits lifted a little. But confusion was soon to follow. At Kenny's Knob, the young German electrician was directed to a hut and handed a hurricane lamp. 'There's no bloody electricity,' the foreman told him when he asked the purpose of the lamp. As the foreman and his vehicle disappeared in a cloud of dust back up the treacherous track, Dieter scratched his head. He was an electrician, yet there was no electricity. So what on earth was he there for?

When he went to inquire about a meal and a shower, he was greeted with another revelation. After all the effort he had been putting into learning English, he suddenly discovered he was now in a camp full of Italians, none of whom could speak either English or German. He walked dispiritedly to his hut, closed the door, fell onto the wire cot which creaked in protest, and cried in the dark. 'What have I done? What have I done?' he asked himself over and over until he drifted into a fitful sleep.

Deep down in the exploratory tunnel the foreman had described, Major Clews's former stock horse handler, Max Paterick, was pondering a similar question. In 1954,

Max and his older brother, Bill, had decided to leave the plains and get work as miners in the exploratory tunnel for the Tumut-1 underground power station. After their meagre earnings as stockmen, the money paid to those who worked in the tunnels was fantastic.

The exploratory tunnel was excavated at a 45-degree angle to a depth of about 300 metres where a chamber the size of a house and, beyond it and through another short tunnel, a second chamber were being excavated. The chambers were the first stage in exploring and securing the foundations of what would grow into a vast cavern to house the power station. The two brothers had teamed up and were working together as part of a four-man shift with two German drillers. Max and his brother worked in one chamber and the two Germans in the other.

When the water-cooled drills were working, the atmosphere became thick with vapour and noise. The only physical senses a man effectively had were sight and touch—and they were of minimal value. Visibility was low; and everything became so wet that even touch became confusing.

One day the drill which the two brothers were operating kept jamming. They wrestled with Stillsons to free it and tried again. Again the drill jammed. After another wrestling match with the big Stillsons, slippery with water and grease, they drilled some more. Again it jammed. The men were frustrated and exhausted. They rested for a moment and Max glanced down towards the floor of the chamber, but all he could see was the dull yellow reflection of his helmet light. He dropped his hand towards the reflection and discovered they were standing waist deep in water. They were so wet from sweat and vapour they hadn't noticed the rising water level.

Max grabbed his brother's arm: 'Hey, have a look here.' He splashed the water. The pair shone their lights around the chamber and saw to their horror that water was squirting in from dozens of fissures in the roof. Max waded to the chamber entrance to find out why the pump was not working. The expended water from the drill and any seepage from the ground were supposed to be drawn away by an electric pump. To his annoyance, Max found the pump attendant fast asleep on the rocky ledge where the tunnel entered the chamber. The floor of the tunnel was about a metre above the floor of the chamber. As Max angrily grabbed the man there was a loud thud and crash behind him. He turned to see his brother's helmet light moving quickly towards him. 'The roof's coming down,' Bill screamed.

'Shit, what about those other blokes?' yelled Max, trying to see if the entrance to the short tunnel that linked with the other chamber had been blocked. The collapsed rock had formed a gleaming black island, which was still growing as rock and earth and water continued to spill from the roof.

Max scrambled over the rockfall, praying that another boulder was not teetering above, waiting to crush him like an insect. When he got to the second chamber the two Germans were still drilling. Isolated by the noise of their work, they had not heard the fall. Max dragged at the men, using sign language to show the roof was coming down. He ushered them in front and they hurriedly made their way back. When they got to the rockfall there was only just enough space left for them to wriggle through. Seconds later and they would have been entombed.

It took two months for a team of miners to reopen and stabilise the chamber sufficiently

TENTATIVE BEGINNINGS

Max Paterick, far right with raised arm, in the Cabramurra wet canteen with workmates.

for work to continue. In the meantime, Max decided to go back to the transport section. Handling horses wasn't so bad after all!

For Max, and others throughout the mountains and on the plains, it was a time of major adjustment. In Cooma the local population, overwhelmed by the pace of change, was beginning to reel in shock. On mountain roads and at construction sites the word 'Achtung' was posted as the new internationally recognised warning sign. It was only a small thing but for Australians, particularly those who had fought in the European war theatre, it was one of the starkest pointers to the changes on the high country.

By the mid-1950s the Authority alone employed more than 1000 professional staff and several thousand workers. It had built three new towns, numerous construction camps, 800 houses as well as messes, hostels, offices, shops, recreation halls and the largest civil engineering laboratories in Australia. It had constructed 160 kilometres of public roads, 176 kilometres of construction roads and 240 kilometres of tracks and had surveyed and mapped 5000 square kilometres of mountain terrain. More than 9500 metres of diamond drilling had been completed.

The Authority had also shifted its head office from Sydney and in addition to East Camp, the home for hundreds of workers, it was building what was virtually a new town, Cooma North. This settlement was for 'staff'. It would ultimately have its own facilities, school, shops and the best houses. Its creation did more to alienate the local community than any other factor, but there simply was not enough land within the existing town area to accommodate the rapidly growing population.

Almost overnight, it seemed, the townspeople of Cooma, who previously knew every face in the street, began to feel like strangers in their own town. Many families in and around Cooma were fifth-generation Australian, direct descendants from original squatters and graziers who moved there in the 1840s. A way of life that had embraced and sustained a community for more than a century was suddenly fragmenting.

Tension and ill feeling had almost reached ignition point when the government revoked high-country grazing leases as a safeguard against erosion and consequent siltation of the new dams and lakes. Summer grazing in the high country had been an integral part of the local economy and ethos since the beginning of settlement there.

When some pet dogs owned by Snowy people were accused of killing sheep, one grazier reacted by throwing strychnine-laced sweets over the front fences of all the houses in the Snowy quarter of town. Two or three dogs were killed almost immediately, and the alarm was raised after one dog was found with the sweet still in its mouth. All personnel were despatched immediately to search their properties for the sweets before any children picked them up. In the end no one was hurt and, although a large number of pets died, there was no noticeable decrease in attacks on sheep.

Resentment also sprang from little things. Shopping, for example, was once a leisurely activity. There was time to pause and chat with the storekeeper and the butcher. Suddenly people had to queue and often they arrived at the counter to find half the shelves empty.

It was as though the town had been taken away from the local people, and they resented it deeply. Not surprisingly they soon began to rebel. The shopkeepers were among the first to take a stand. They served the locals first and chatted amiably, pointedly ignoring the long queues of Snowy people who eventually took the hint and left. Sometimes a Snowy person would walk into an empty shop only to be told by the store keeper that he was too busy to serve him—he was making up the country deliveries.

Even the milkman dug in his heels and refused to deliver to the Cooma North houses. The resulting antagonism was resolved only when the Cooma North residents agreed to hang their billies, containing the money, on nails hammered into the wooden streetside power poles. The question of where the milkman would leave any change that might be due was soon rudely answered—it was in the milk.

The town's social order was overturned, with senior engineers toppling landowners from the upper rungs of the social ladder. One early counter to this was launched by the local golf club, which passed a law barring from its board anyone who had not been a resident before 1949.

In the schoolyard Australian children who had little concept of a world beyond the Monaro, were suddenly confronted by children who appeared to them like creatures from another planet. Ian Mould, whose great-great-grandfather rode on horseback into the district in 1850 to become the Monaro's first resident doctor, could scarcely believe his eyes the day the first migrant children turned up at the little bush school.

On a day which started like any other at the Cooma school but which would change everybody's lives, the scuffles, cricket and hopscotch on the dusty, clay playground were halted and the children were assembled before the headmaster. With him were three strange boys who stared impassively back at the gawking assembly. He introduced them as three classmates from Germany, Estonia and Holland. Germany and Holland the assembled students knew about. Australia had fought wars against Germany; and Holland was a flat place with windmills where a boy was supposed to have stuck his finger in a hole in a dyke to hold back the sea. But Estonia? Where was that?

The children in the playground began to giggle and jostle. The three foreigners wore strange-looking leather breeches over brightly coloured shirts and had heavy-looking shoes and long bright-coloured socks. One of them even wore a funny little hat. Australian boys wore only simple cotton shorts and shirts and only in tones of grey and brown. Anything else was sissy.

The assembly was dismissed and the Australians scattered to resume their activities in the precious minutes remaining before the bell summoned them into class. A few ventured across to the strangers to get a closer look and ask why they were there. But it was obvious after a few words that they could not even speak English; so the curious wandered off to find their mates. The strangers were left in each other's company, united by their alienation.

Lunchtime produced further differences. The Australians took sandwiches of white bread with cold lamb and tomato sauce. It rarely varied. The newcomers, however, tucked enthusiastically into black bread and cheese, sausage and garlic. The Australian children looked on with mingled awe and disgust. They had never before encountered garlic. Its pungent odour repulsed them and the migrant children were quickly dubbed 'garlic munchers'. The newcomers, for their part, could not see what all the fuss was about and regarded the Australians as being eccentric, even backward.

As the weeks passed, more strange children, in a range of ethnic costumes and speaking a variety of tongues, arrived in ones and twos until they were no longer a matter of curiosity to the Australian children. It usually took a few weeks for the migrant children to settle down. The Australians were distrustful and sometimes cruel. The newcomers were frequently baited and ridiculed and kept on the outside of friendship circles.

But in time the barriers did break down. Once the migrant children had a reasonable grasp of English, the teachers would get them to stand in front of the class and talk about their home countries and communities and explain their customs and home life. In this way, the Australians began to learn of the hardships many of the new children had endured in postwar Europe.

But this did not necessarily lead to any sudden expression of compassion or understanding. Ian Mould remembered one Dutch family who had arrived in Cooma with nothing but a few suitcases: 'We thought it was quite funny. We used to laugh at the kids' thick-soled shoes which their dad had made from old car tyres.'

But before long the migrant children were so numerous that they began to exert some influence over the established school routines. School life in a small Australian bush town like Cooma was simple. In summer it revolved around cricket for the boys, and softball, rounders, or hopscotch for the girls; in winter it was rugby and netball. None of these games made any sense to the newcomers. As winter settled over the Monaro, the migrant boys wanted to play soccer, a development the headmaster thought would be good for everybody. Before long the Australians found themselves surlily kicking a soccer ball. They did not know the rules and suddenly the foreigners were in command.

In the succeeding months and years a transformation overtook the schoolyard. The migrant children learned English quickly, began wearing the same clothing styles as the Australians and some even started playing cricket. Gradually, they became just another element of the local way of life, until it was difficult to remember what it had been like before their arrival. The Australians, too, changed. In time it was no longer 'sissy' to wear more colourful clothes, particularly in summer. They also learned that 'wog food' wasn't so bad either. Some, like Ian Mould, even granted the 'new kids on the block' a grudging respect: 'The foreign kids adapted a darn sight faster than we would have. By the time I left school—two years after the first migrant kids had arrived—some of them were dux of their classes. Two years earlier they hadn't even been able to speak English.'

A similar change was evident in the adult world. Cooma was being changed and broadened into a community that was more cosmopolitan than any Australian city at that time. Brothels, 24-hour hotels and nightclubs were turning once-quiet Cooma into a frontier town. A big 'boarding house', which at different times had been a hospital and a grammar school, was now doing thriving business as a bordello. To the locals, who preferred not to acknowledge its nocturnal activities, it remained a boarding house. But the Snowy workers knew it as 'Maria's Place' and, judging by the long line of cars outside every night, Maria was a popular woman.

Other changes were reshaping the town. The cafes, whose menus were limited to bacon and eggs, ham and eggs, sausages and eggs, mixed grill or fish and chips, were suddenly in competition with restaurants offering international cuisine—Chinese, French, Italian, Austrian and others. The locals still labelled it 'wog food', but they ventured into such establishments with increasing regularity.

Despite the changes Cooma still offered sights and sounds that left the Europeans transfixed in wonder. Friday was still sale day. Drovers on horseback, cracking stockwhips and yelling unintelligible commands to darting, barking dogs, still brought sheep to town from off the plain; all day Friday, mob after mob, crossed from one side of town to the other. It was an incredible sight for people raised in the cities of Europe.

There were also fascinating new sights for the local children as the momentum of the Snowy Scheme began to pick up. 'None of us had ever seen so many vehicles and big trucks,' recalled Ian Mould:

> As kids we'd spend ages just watching trucks drive by and studying the imprint of their big tyres on the dirt roads. Occasionally there would be real excitement when one came by with a house on its back. We'd never seen that before.
>
> There wasn't much in the way of entertainment. We'd spend a lot of time in the bush, playing war games and the usual kid stuff. The picture theatre would open on Friday night and there'd be a Saturday morning matinee for kids. You never had to book, just turn up.
>
> But that changed with the Snowy. You would turn up as normal and find that because you hadn't booked a seat you couldn't get in. It was our theatre and we couldn't get in. It was that sort of thing which sparked resentment.
>
> There was one spin-off though—the quality and age of the films screened improved out of sight.

Before the Snowy there was no night life in Cooma except a dance three or four times a year in the Country Women's Association hall and at the agricultural and pastoral show. With the Snowy came regular Saturday night dancing. But it was tough for a bloke to find a partner. There were hundreds of eager young men and just a handful of young, single women, most of them nurses from the local hospital. Most parents of local girls strictly forbade contact between their daughters and the migrant men.

Eventually, out of an unspoken belief in giving a person a 'fair go', the locals accepted migrants as individuals. The first workers to arrive had been displaced persons. When they had arrived, hollow-cheeked and devoid of possessions, they had to be taken to the general store by Authority staff to be kitted out with such basic needs as clothes, dinner plate and a knife and fork. The people who witnessed these poignant transactions were deeply moved, but in no way did this diminish the broader resentment against the Snowy 'invasion' and the inconveniences it caused.

One of Ian Mould's Saturday morning chores was to take his billycart into town and drag home the shopping: 'I often used to be able to do it in twenty minutes. But once the Snowy arrived I'd find myself waiting in a queue of twenty or thirty people.'

One of the reasons the local shopkeepers were reluctant to expand their operations to cater for the increasing population was that they, like everybody else, knew the new government was stalling in declaring its full support for the Scheme.

The man who bore the brunt of this political oscillation, as well as the other problems confronting the project, was the commissioner, Bill Hudson. He tried to counter opposition by mounting a vigorous public relations campaign, but it made slow progress. When the Scheme was a year old he invited 200 state and federal parliamentarians to visit the region to gain some understanding of the scale and scope of the planned projects. Twelve turned up.

Even when, in 1951, the federal government agreed to a proposal to call on the expertise of engineers from the United States Bureau of Reclamation and to send Australian engineers to America for advanced training in large-scale construction, the agreement contained an escape clause. Cabinet pointedly stated: 'The decision involves no prejudgement of the extent to which any section of the Snowy works are to be proceeded with when works priorities, generally, are under review.' Such indecision only served to spur Hudson on. He pushed his staff mercilessly to achieve results and make progress. Despite the obvious political pressures, some felt he was moving too fast and going about things the wrong way. It was not long before Hudson had to watch his back as well as attend to the endless details of the work.

One of his associate commissioners, Thomas Lang, disapproved strongly of Hudson's style. Lang was a brilliant but methodical engineer, who often believed the pace demanded by Hudson would lead to costly mistakes. He was proved correct on one project. After the expenditure of more than a million pounds on Adaminaby Dam, the site chosen was proved to be inadequate, and the dam had to be shifted.

That episode gave both the Authority and Senator William Spooner, the government minister responsible for the Scheme, a serious fright. Hudson had to decide whether to make the change and admit to Spooner that the Authority had wasted a lot of money; or play safe by settling for what had already been started, even though it would mean a much smaller water storage capacity. Hudson had the courage to admit the mistake.

Spooner took the news calmly and prepared a press statement for release on his favoured day, Saturday. However, his calm was shattered when on the Friday afternoon a finance scandal rocked the New South Wales government. A state department had attempted to cover up the mis-spending of an amount that was miniscule compared with what had been wasted on the Adaminaby Dam; and heads were rolling. Spooner also had the galling task of informing Menzies of the waste of funds. Spooner was a keen supporter of the Scheme, but he was unsure sometimes which way Menzies would jump.

Spooner and Hudson discussed the problem late into the night. The temptation to try to hide the matter from public eyes was strong. Finally, however, they decided to announce the change of plans, admit but play down the wastage, and arm themselves with a detailed explanation of the reasons for the change and the benefits and savings that would accrue in the long term.

Senator William Spooner (left) with an Australian-born foreman Bill Simpson (second from right) and German-born tradesmen at Adaminaby, 1950.

Jounama Pondage where the water released by the Tumut–3 power station is collected. From here water can be pumped back up into Talbingo Reservoir for recycling through the power station before release further downstream to Blowering Reservoir.

Over page

The kilometre-long vehicular entrance to the Tumut–2 underground power station.

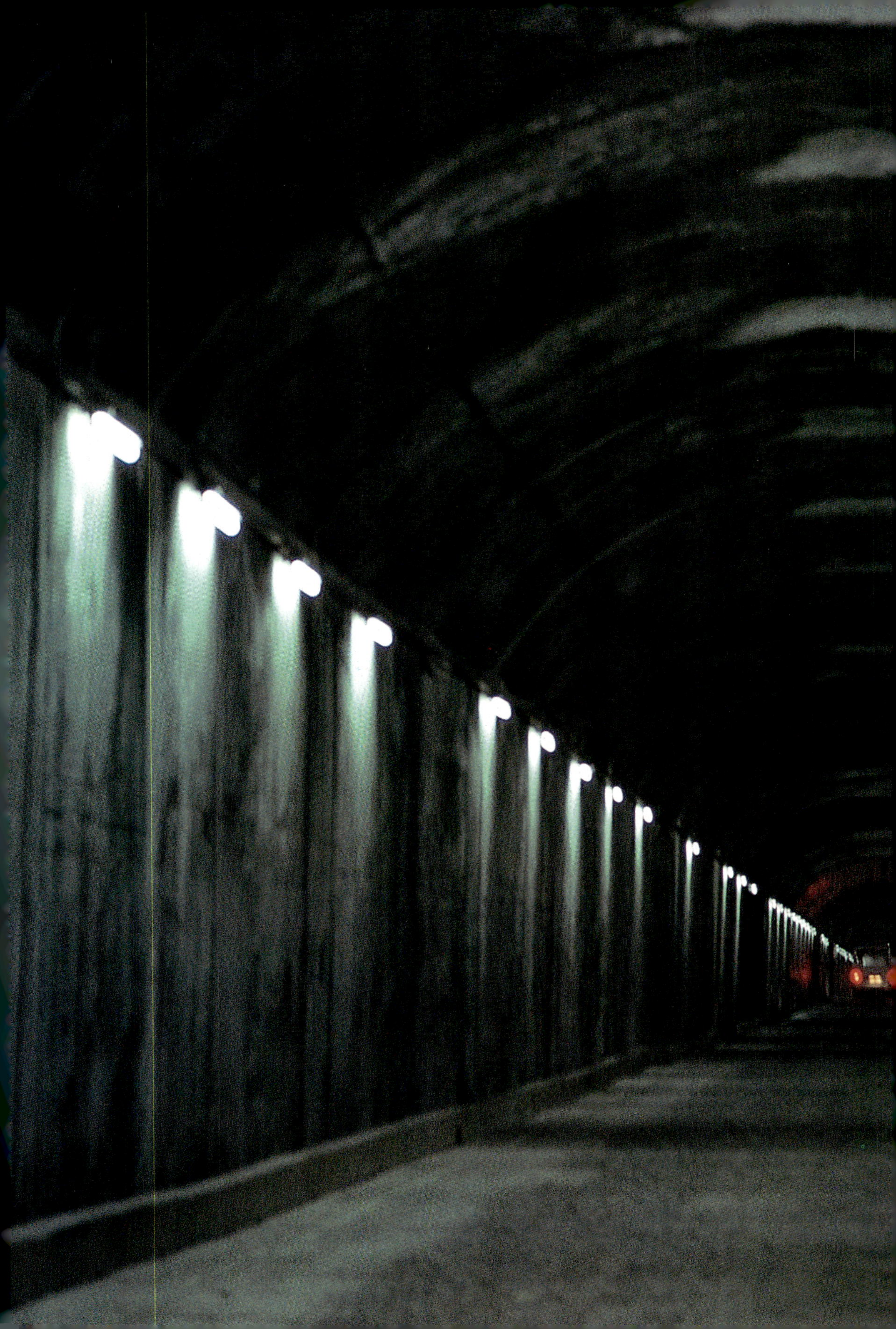

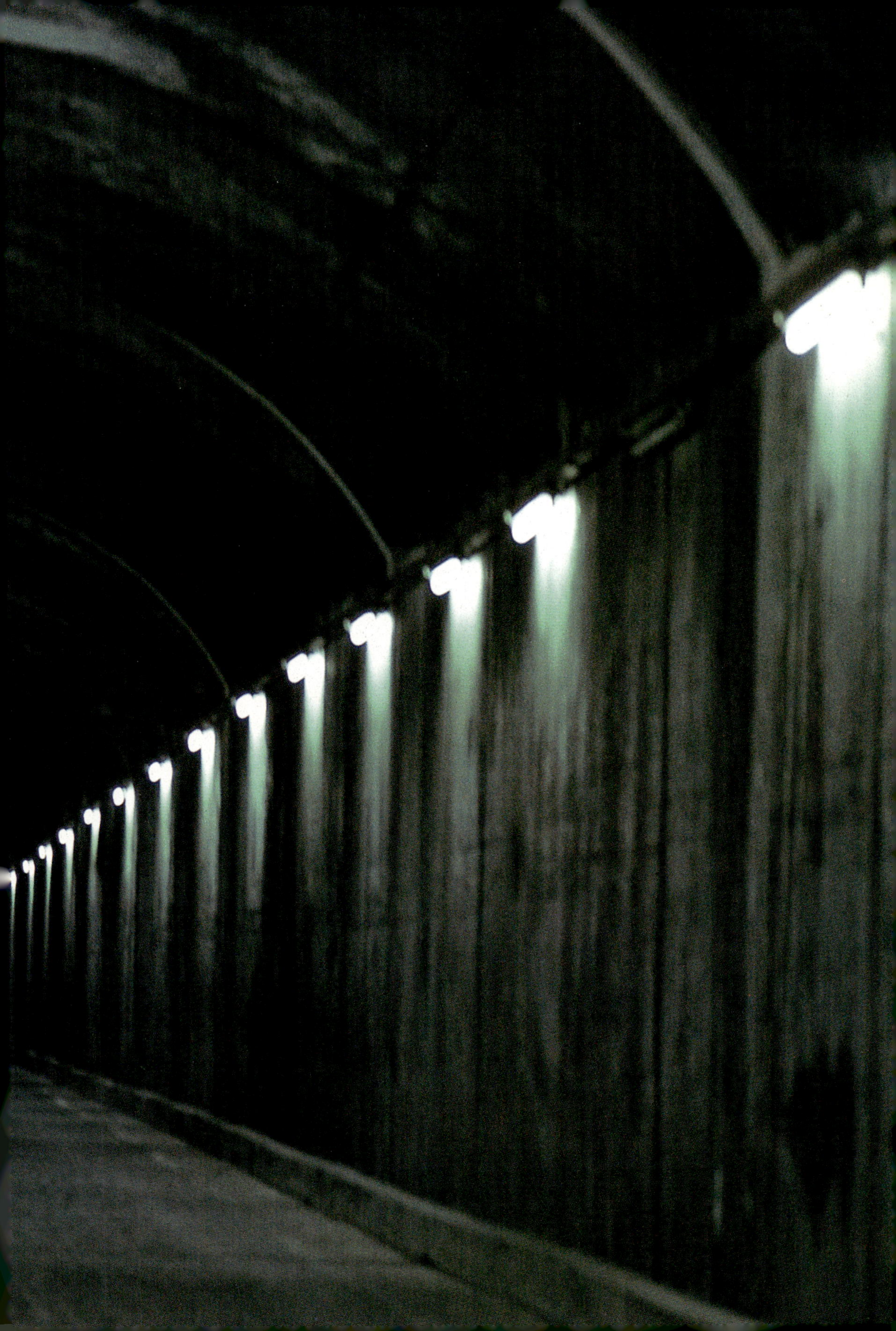

Talbingo Reservoir, the head storage for the Tumut-3 power station.

Murray-1 power station.

TENTATIVE BEGINNINGS

Thomas Arthur Lang, Sir William Hudson and Tony Merrigan when Hudson was knighted, June 1955.

The statement was released and the two men waited nervously, ready to tough out the reaction. But the statement was virtually ignored. The only public comment was in a *Sydney Morning Herald* editorial which contrasted the attitude of the Authority with that of the state department and congratulated the Authority for its frankness which, it suggested, was a model for other government bodies. The incident greatly strengthened the bond between Spooner and Hudson who became an increasingly influential team.

The outcome may also have put paid to Lang's ambitions. From the beginning Lang had expected that he would eventually take over from Hudson. Lang was only thirty-five when he was appointed associate commissioner, and many young graduate engineers from Melbourne and Sydney universities had followed his example and accepted work on the Snowy instead of seeking overseas experience.

Little has been said of the internal power struggle between Lang and Hudson. It came to a head in 1955, when Lang supposedly took his grievances against the Hudson style to Senator Spooner. Spooner in turn passed the report on to Hudson for his comments. Lang later resigned and went to California, where he achieved distinction as one of the most highly regarded civil engineers in the United States.

In his seven years on the Scheme Lang made a crucial contribution. He is credited with being one of the key men behind the development of rock-bolting, a technique developed by Snowy engineers to bind and secure unstable subterranean rock formations. A horseshoe pattern of holes was drilled into the tunnel walls and roof and long steel rods with expansion heads were inserted and tightened against a bearing plate on the outside face. The method turned potentially unstable rock around a tunnel into a solid self-supporting arch. It was faster and cheaper than constructing steel frames or lining with concrete. It was acclaimed by engineers around the world and is now a standard tunnelling practice. Lang was also the driving force behind the Authority's decision to approach the United States Bureau of Reclamation for expert advice on dam construction and on the management of large contracts. Australian engineers in the 1950s had simply no experience in managing projects as vast as the Snowy.

One of the first pieces of advice the Americans gave was to abandon the sole use of single-shift day labour, a traditional Australian work practice that was already being used at Adaminaby Dam and on the Guthega project. They also recommended the

Guthega power station under construction at the confluence of the Snowy River and Munyang River (right).

TENTATIVE BEGINNINGS

Early work on the Adaminaby Dam site on the Eucumbene River by the NSW Public Works Department, 1954. The dam was later moved further downstream and its name changed to Eucumbene Dam.

Authority move away from the traditional cost-plus basis for tendering, and switch to contracts based on target estimates and fixed fees. All this was new to Australian industry. Before the Snowy, everything had been based on day labour, and there had been minimal concern with questions of job duration and worker productivity.

On the advice of the Bureau of Reclamation, the Authority instructed intending contractors to tender an estimated cost. There was an agreement that any savings below the tendered figure would be shared between the contractor and the Authority. The Authority also set time schedules, with bonuses to the contractor for finishing on time and penalties for late completion. The Bureau also entered into an agreement to train 110 Australian engineers in groups of twelve over a ten year period, beginning in 1952. To help Australian engineers gain confidence in handling projects alone, the Bureau sent out a number of specialists in contracts administration to oversee the Authority's early arrangements.

The first major thrust of the Scheme, apart from the small stand-alone Guthega project, was the design of the Upper Tumut development. This comprised, initially, the Adaminaby Dam (Lake Eucumbene); a 22-kilometre tunnel from this dam to Tumut

Pond; a dam at Tumut Pond; and an underground power station, Tumut-1. The design was drawn up in the Bureau of Reclamation's Denver office, with their senior staff using Snowy engineers as junior assistants.

The first contracts signed under the new system were with a French consortium, led by Etudes et Entreprises, for Tumut-1 underground power station, and with an American consortium, Kaiser-Walsh-Perini-Raymond, for the 22-kilometre Eucumbene–Tumut tunnel to Tumut Pond Dam, the headwater for the power station. The dam would be a concrete arch wall eighty-six metres high and 218 metres across the top.

One of the first Snowy engineers to train with the Bureau was Ken Andrews. Andrews was given the job of supervising the contracts, with the Americans looking over his shoulder to make sure 'I didn't put my foot too often in the wrong place. They were a great help in heading off mistakes before we made them. But after three or four years they felt we knew the ropes well enough for them to leave us.'

The Australians soon developed enough confidence to depart from the American way and develop their own ideas. 'The Americans were very conservative in their designs. They designed the Eucumbene–Tumut tunnel to be concrete-lined along its entire length,' said Ken Andrews. 'But I'd gone in with the tunnellers and examined the rock and reckoned it would last till Kingdom Come; so I figured why line with concrete when the concrete was no better than the rock?'

Andrews and the other engineers decided instead to line only those sections that warranted it. Ultimately only twenty per cent needed lining, which saved the Authority a considerable sum of money.

> We used our heads, rather than slavishly follow everything the Americans did. This was one of the reasons why the Snowy Scheme was built within its budget estimate. We gained the confidence to use commonsense and tailor designs to the circumstances.
>
> There were also time bonuses for contractors and workers for getting projects finished early. That lowered the cost considerably because it meant we paid less interest on our loans.

In those early years both contractors and the Authority's engineers worked long hours, pushed relentlessly by Hudson, who sometimes acted as if there would be no tomorrow:

> We all admired 'the old man'—sometimes, of course, we hated him but even then, you admired him. I lived just down the hill from him and, working a hundred hours a week, I didn't get to bed until about one o'clock in the morning. Up the hill, though, Hudson's study light would still be burning long after mine went out. You felt guilty about going to bed at all.
>
> He set standards which you felt obliged to follow. One Friday afternoon he called all the senior engineers into his office for a conference. We were still there at four o'clock on the Sunday afternoon when the commissioner suddenly looked at his watch and exclaimed: 'My goodness, it's getting a bit late. You blokes better go home, I don't want to spoil your weekend.'
>
> He was serious and didn't realise how funny it was.

TENTATIVE BEGINNINGS

Eucumbene Dam, 1959, a year after its completion created Lake Eucumbene as the Scheme's main water storage.

The American contractor charged with the Eucumbene–Tumut tunnel and Tumut Pond Dam set a blistering pace, pushing its workers with unremitting vigour. It had divided its workforce into three shifts to enable the work to proceed twenty-four hours a day, six days a week. Sunday was the only day off and even that was only to allow maintenance teams to overhaul equipment.

The Americans demanded total effort from every man. They had their eyes set firmly on the bonuses offered by the Authority for getting the contract completed early. They developed a reputation for ruthlessness and fatalities started to occur as tunnelling crews, in particular, were pushed harder and harder until they were regularly setting new world tunnelling records. Three deaths for every mile of tunnel was considered acceptable by the Americans. As they were keeping just under this, they imagined themselves beyond reproach. To many of them, the migrant workers were no more than well-paid, and therefore expensive, slaves.

Quite apart from the safety issue, the rapid progress being made by the Americans soon presented the Authority with yet another political headache.

In order to buy peace with New South Wales, the Commonwealth had to give the construction of the Adaminaby Dam to the New South Wales Public Works Department. However, the dam construction fell further and further behind and it began to look as though the tunnel from Eucumbene to Tumut Pond would be finished and then the whole project would come to a standstill—perhaps for several years—as it waited for the Department to catch up. To Hudson it was a terrifying prospect.

He raised the matter on several occasions with Senator Spooner. In May 1956 Spooner urged federal Cabinet to put pressure on the New South Wales government to relinquish the contract and call tenders to finish the dam.

The Public Works Department had built a wonderful work camp, comprising about 400 houses, sealed streets, a shopping centre and a town hall that was the envy of every other community in the region. As far as the Authority was concerned, the Department had been too busy settling in to do the work it had been contracted to do—namely build a dam.

The federal pressure wasn't heeded, and Hudson finally decided to take direct action. The Department, effectively, was sacked and the Authority called for tenders to complete the work. The tunnel contractor Kaiser-Walsh-Perini-Raymond won the contract and in two years completed more than the Department had achieved in four.

In their own defence, the Department's engineers cited as one of their chief problems their lack of control over the labour force. The project moved according to the unions' wishes, rather than theirs. When the Americans assumed responsibilty for the Adaminaby Dam the unions engaged immediately in a test of strength and called a strike over the new work practices the Americans intended to introduce. The Americans responded by sacking the entire workforce and shutting down the project. After a week the rank-and-file workers began to return, seeking their jobs back.

The change which took place under the Americans was far-reaching. They introduced modern messes and contract caterers, and revolutionised employment terms, conditions and work expectations. To this day there are sections of the labour movement in New South Wales which have neither forgotten nor forgiven the Authority for sacking a public labour workforce. In other people's eyes, however, the labour movement had been setting up the Snowy as a 'jobs for life' project.

About this time—and probably not coincidentally—the Scheme began to suffer its first serious public attacks. Articles started appearing about wastage of materials and newspaper reports, extensively quoting unnamed 'experts', claimed the Authority's dams were nothing more than inspired guesswork. When planning for the Scheme, the report claimed, the Authority had relied solely on data obtained from a single river gauge. Rather than respond directly, Hudson issued another invitation to federal and state politicians to visit the Scheme. This time thirty turned up.

Most of the workers who returned to work, thus breaking the strike, were migrants who were suspicious of unions. They were on the Scheme to work and earn high wages, not to play politics. They subsequently made it their business to keep militant unionists or troublemakers off the site.

The unions had only themselves to blame for this reaction. Many migrant tradesmen were angered by Australian trade union attempts to discredit overseas qualifications and by the union push to have tradesmen from Europe start afresh and requalify. A number of qualified tradesmen, working as labourers, became so frustrated that they returned to Europe as soon as their two-year work contracts expired.

Almost from the beginning, the growing workforce on the Scheme had attracted the attention of about a dozen trade unions. The biggest and most powerful of these was the Australian Workers Union (AWU).

At first the unions found it difficult to sell membership to the migrants. The Germans, in particular, were antagonistic and there was trouble between them and the pro-union Australian and British workers who were trying to coerce all workers to join up.

To avoid a major industrial dispute the Authority and contractors arbitrarily agreed to deduct men's union dues from their wages. This added further to the discontent— particularly of the Germans, who had come out as skilled tradesmen and were already unhappy with the working and living conditions. For some, the conditions were far worse than they had expected. A few actually walked off the Scheme. This prompted the Minister for Immigration, Harold Holt, to issue a statement in April 1952 warning that migrants who left their work on the Snowy were in breach of contract with the Australian government and would be tracked down and deported.

Holt's warning came after three German tradesmen, a mechanic and two fitters, one day refused a directive to transfer from Island Bend to Perisher. They complained that Perisher had no mess and that the only accommodation was tents with no heating. They also complained that they were expected to work in snow and rain without rubber boots or raincoats. The three men actually demanded to know if Australia had signed the United Nations Charter of Human Rights, claiming that the conditions under which they were expected to live and work in Australia were worse than those they had endured in Russia during the war.

Such statements hardly endeared them to the Authority, which denied the claim about the rubber boots and raincoats and added that the harsh conditions had been fully explained to the men before they had signed on in Germany. The engineer, Walter Hartwig believed most of the trouble among the Germans was caused by misunderstandings that resulted from the language barrier:

> They were tradesmen with little or no English. They had been landed in Melbourne, put on a train to Cooma and then sent out to primitive camps where no one could understand them.
>
> The first group was sent to Smiggin Holes. It was winter and it was snowing. Apart from those who fought on the Russian front, the men had never experienced such conditions before.
>
> From the very beginning when I was helping the Authority to recruit in Germany, I had reservations about employing city men. I couldn't see cultured Berliners coping with the Australian bush—though eventually most of them proved me wrong and they did.

In the end the Authority let the three main antagonists return to Germany and, because it feared the consequences of a legal challenge to its constitutionality, it did not prosecute for reimbursement of their fare from Germany. The extreme feelings of the three men were not considered to be representative of other workers. Also, on a project the size of the Snowy some industrial disputes were thought to be inevitable.

The first and biggest had occurred in the summer of 1952 when the Authority, under orders from Canberra to cut costs to the minimum, withdrew the workers' bus service. From the beginning of the construction program the Authority had used buses to ferry workers from the East Camp barracks to various work sites around Cooma. At the end of 1952 it told the workers they would henceforth have to organise their own transport.

The workers, rallied by a core group of Yugoslavs and Australians acting under the auspices of the Australian Workers Union, immediately went on strike. When a meeting of all affected workers was called, Bill Hudson despatched a spy to find out what action the men intended to take. After the meeting dispersed, the spy ran ashen-faced to the head office to report the workers had voted to burn some of the Authority's buildings 'to teach the bastards a lesson'. Hudson did not really believe they would be that militant, or so foolhardy, but as a precautionary measure he ordered all divisional chiefs to post twenty-four-hour guards over buildings under their control.

All buildings, including the laboratories of the Scientific Services Division at the other end of town, were of weatherboard. Because of the Division's isolation from the rest of the Authority, it was considered particularly vulnerable.

A roster was drawn up and hydraulics specialist Jack Lawson with a New Zealand engineer, Noel Carter, were given a midnight-to-8 a.m. guard shift over one of the main huts that was used for model-building. When the pair turned up shortly before midnight to relieve their senior engineer, Wally Shellshear, they were alarmed to find him armed with a loaded rifle. It gave a whole new dimension to the situation, suggesting a seriousness they had not before appreciated.

The two young engineers took over the watch and sat on the floor in the centre of the building, talking, playing cards—anything to take their minds off the peril they now felt themselves to be in. The workers, particularly the Yugoslavs, were a rough lot; their violence after a few drinks was already well known. They had not the slightest idea what they were expected to do should a mob turn up and torch the building. There was no telephone and, worse still, only one exit, which they had bolted. It occurred to them that they faced the very real prospect of being fried alive.

About two o'clock they saw, through the hut's fixed, unopenable window, the town's street lights turn off. A black silence descended on the world outside. At three o'clock their spines rippled at the sound of a truck engine in the distance. Their senses heightened by their growing fear, they strained their ears to follow the noise and realised it was coming their way. They could see the headlights lighting up the trees as the vehicle climbed the hill leading to the complex. Noel Carter pulled the light cord that hung from the roof and the two men sat fearfully in the dark.

The vehicle slowed. Its tyres crunched the gravel as it turned into the yard outside, its headlights momentarily flashing inside the hut as they crossed the window. The motor stopped and sounds that froze their hearts carried to them easily in the still night air—the rumble of men's voices and the clanking of metal containers, which they had no doubt would be brimming with petrol. There was more clanking, accompanied by muffled talk and the grunted breathing that goes with exertion. Then, abruptly, the truck engine started and the cab doors slammed. As the truck began to move off the pair walked nervously to the window to see Cooma's night-cart operators driving off. They had been replacing the pans in the outside toilets!

Neither man said anything for a while. Then one of them sheepishly pulled the light back on, and they made a pact never to mention the episode again. It later transpired that the strike organisers, fully expecting that the Authority would have staff monitoring the meeting, had threatened the fires to 'give the buggers something to think about'.

In the strike's third week, the workers were assembled in the recreation hall. They were restless and wanting to go back to work. Because they were on strike, the mess was closed; and because they were earning no money, they could not buy food. Consequently they were all very hungry. Union officials from Sydney, who had now become involved, expected this and had representatives move through the crowd doling out fresh tomatoes. They clearly had not dealt with European workers before. When the union leaders stepped into a boxing ring in the centre of the hall to address the men and bring them up to date with negotiations, it rained tomatoes. The Europeans then strode out and returned to work. Hudson, seeing them defy the union, immediately reintroduced the buses.

Although the strike lasted three weeks, it proved but a minor episode in the increasingly intractable legal problems that confronted the Authority's industrial procedures. The Snowy Mountains Hydro-Electric Power Act, which established the Authority, made no provisions for the machinery and procedures necessary for handling industrial disputes. The Authority could not approach a state industrial tribunal because it was a commonwealth body. Conversely, it could not deal with the Commonwealth Conciliation and Arbitration Commission because its activities were confined to New South Wales and the commission's role was the settlement of inter-state disputes.

The Authority desperately wanted to apply for a special award to cover its workforce under the Commonwealth Court, but this was fiercely opposed by the unions. State awards at the time embraced far superior pay and working conditions than did federal awards. Every move by the Authority to seek federal coverage prompted a warning from the unions, particularly the AWU, that they would challenge in the High Court any industrial provisions granted. While it was never stated as an open threat, everybody knew that such action would automatically become a challenge to the validity of the Act governing the Authority and, by implication, the Scheme itself.

The issue dragged on for years, with the Authority and the federal government sinking steadily deeper into a legal quicksand as a result of their attempts to outmanoeuvre the states—particularly the recalcitrant New South Wales government.

At the heart of their dilemma was the way in which the Chifley government had overridden state rights in legislating for the Scheme in the first place. Constitutional validity for the Act was based on the Commonwealth's powers to make laws with respect to defence—in this case, the desirability of providing electricity for war, plus reliable power for the Australian Capital Territory, the seat of government. There was and still is nothing in the Constitution to prevent the Commonwealth from using rivers as it saw, or sees, fit for defence purposes. But four years had elapsed since World War II and Australia had abundant coal for its thermal power stations.

Even in the very beginning Prime Minister Chifley confided his unease, telling Cabinet that he felt the defence powers provided a 'dubious foundation'. He urged every effort to persuade New South Wales and Victoria to validate the Scheme in their state parliaments as soon as possible. 'It would be a bold legal adviser who asserted there was no element of doubt as to the prospects of a High Court judgement upholding the basis of defence powers in a time of peace,' Chifley said.

The defence argument, however, continued to be pushed in various guises for some time. In 1954, in an address to the Australian Institute of Political Science, Hudson re-emphasised the defence aspects of the Scheme, saying that one of the reasons for having two power stations underground was to protect them from air attack. The real reasons for siting the power stations underground were the rugged topography and geological considerations.

With the change of government, when the Scheme was just six months old, the Menzies Cabinet inherited the legal headaches in which the project was embroiled.

In October 1950 Menzies tried to stall the states by suggesting that, as an interim arrangement, they settle merely on an agreement for the distribution of water and electricity from the Scheme.

The Solicitor-General's office had advised Cabinet that if the Act on which the Scheme was based was eventually declared unconstitutional, the Commonwealth could then invite the states to join with it in providing the necessary legal basis to salvage a great national undertaking. 'By that time the present ambitions and resentments may be less keen and the work will be well advanced,' the Cabinet minute read.

In November, a Cabinet meeting, attended by Hudson, decided that the Attorney-General should prepare and circulate to the two states a draft agreement for ratification by their parliaments. The Attorney-General was also asked to have his department prepare a draft bill, which the states could consider, giving the Snowy Authority control over the Scheme.

The states, of course, were unlikely to accept any such 'suggested draft' from the

Commonwealth, but the tactic bought the Commonwealth more time while the offer was discussed. It also permitted the Commonwealth to be seen as trying hard to accommodate the states. In the meantime, until an agreement was ratified by the states, federal Cabinet was adamant that the Snowy Authority must avoid appearing in any courtroom, no matter what the issue. Any court could be the starting point for a challenge to the Scheme's constitutional validity.

The states accepted the interim agreement for water and power distribution. But the Commonwealth's attempted subterfuge, far from pulling it from the mire, pushed it in even deeper. The state administrators were not as easily duped as some federal bureaucrats might have wished.

The main plank in the Commonwealth's defence against any constitutional challenge was the argument that it needed to build up a reserve of generating capacity for war. Australia had a reserve of thirty-five to forty per cent in 1939, which had diminished to a negative figure by 1945—hence the blackouts which were a regular feature of postwar life in Sydney and Melbourne. The Commonwealth would argue, therefore, that to start a long war with meagre electricity reserves would severely curtail war production.

But none of this was mentioned in the interim agreement, which was merely an arrangement whereby construction of the Snowy would proceed at a rate which would supply power in accordance with the general growth of the New South Wales and Victorian electricity systems. Put simply, it was an agreement by which normal peacetime growth in electricity demand would be jointly met by the Snowy and the states' thermal power stations.

Any mention of the defence aspects of the Scheme would have caused the states to point out that provision for war was a Commonwealth responsibility. This being the case, they would have insisted the Commonwealth supply the Snowy power at an appropriately subsidised price. The Commonwealth could not afford this because the financing arrangements for the Scheme depended on the states' paying full-cost rates for the hydro-electricity.

The federal government had painted itself into a sticky corner. Its only recourse was to continue to try to bluff its way out while keeping the Authority out of court. Until the preparatory work was completed and major works well underway, New South Wales might still seek to thwart the Scheme by legal means and implement its long-favoured irrigation-only proposal. Underlying this situation was also the principle, widely accepted at the time, that the provision of power and water was a state, and not a federal, function. Federation was less than fifty years old, and all states were sensitive to any perceived encroachment on their rights.

The federal government's fear of a legal challenge to the Scheme hogtied the Authority in all matters of law. It could not use its power to resume land for fear of being taken to court on the constitutional issue. It had to acquire land by negotiation with the owners, rather than by use of the Commonwealth's compulsory powers. Nor could it sue for rent due on Authority-owned houses, or secure the eviction of those no longer eligible to occupy the houses. It could not even sue for the return of stolen money.

The unions were well aware of the Authority's position and used it to their advantage. But even they were constrained from using the constitutional doubts as more than a bargaining threat. The project was, after all, providing thousands of jobs which the unions did not want to lose.

The eventual solution evolved almost of its own accord. It was unique in the history of industrial relations management in Australia, and provided some valuable (but in the long term unheeded) lessons.

The Scheme was the biggest construction project in Australia's history yet, apart from the three-week strike in 1952, there were no other employee stoppages longer than an hour or two for the remaining twenty-one years of work. The entire workforce expended maximum effort to achieve a remarkably high rate of production and the Scheme was completed on time and within budget.

Two basic factors contributed to this. The first was the decision to appoint a single arbitration commissioner to specialise in the Scheme. Initially this was the Honourable Mr Justice S.C.G. Wright, of the Commonwealth Arbitration Court. However, the unions opposed the involvement of a federal court, which carried with it the risk that federal awards might replace the preferred state awards.

Hence the second factor—the appointment of Mr Justice S.A. 'Stan' Taylor, president of the New South Wales Industrial Commission. Judge Taylor simply grew to love being in the mountains among the migrant workers. He formed deep friendships among both management and workers—in particular with Charlie Oliver, secretary of the New South Wales branch of the Australian Workers Union. Taylor devoted himself to the Scheme, spending more and more time in the mountains and less and less time in his chambers in Sydney. He and Charlie Oliver became an informal conciliation and arbitration commission, solving disputes before they flared. He was generous to the workers but rigorously upheld management's right to expect maximum effort from its employees.

Despite Taylor's moderating influence, the Authority continued to be plagued by problems and anomalies caused by the mixture of state and federal awards under which workers operated. The Authority adopted New South Wales awards, but was far from happy with the arrangement. The cost of labour would be considerably cheaper under federal awards and Hudson was a stickler for keeping within budgets.

The strong union resistance that had greeted the appointment of the federal arbitration judge, Mr Justice Wright, in 1952, did not deter the Authority from trying again, in 1955, to have the workforce brought under a single federal award.

The Australian Workers Union promptly filed an application in the High Court, asking it to declare that any such industrial provisions added to the Snowy Mountains Hydro-Electric Power Act could not be justified under the defence provisions in the Constitution. It argued that under the Act the workforce, technically, was engaged in defence work. The Commonwealth Arbitration Court, therefore, could not impose a civil industrial law on the workers. But the claim was never tested. The union's High Court application alone was enough to prevent the Commonwealth Arbitration Court from proceeding with an award.

The growing frustration within both the Authority and the government was reflected in a strident press statement by Senator Spooner, in which he denounced the union's High Court move:

The Hon. W.H. Spooner.
PHOTO: NATIONAL LIBRARY CANBERRA

There must be no danger of power shortages if Australia is again called upon to defend itself and to shoulder its responsibilities to its allies.

Anyone, or any group, impeding the progress of the Scheme in any way is doing deliberate disservice to Australia.

Patriotism is a matter of action, not words.

The draft of the statement had an additional sentence which was removed on the advice of departmental lawyers. It read: 'Whatever be the motive or intention of the throwing of this spanner in the works is sabotage of Australian defence at home in Australia.'

The early 1950s was a time of mounting paranoia about Communism. It was a popular weapon of the government to attack trade unions with the slur that they harboured Communist ideals, even active and dangerous Communist cells.

Senator Spooner sent a copy of his statement to Hudson, to which he attached the following plaintive note:

I decided to make a press statement but it was not published by the papers. The difficulty is preparing a statement which has news value but does not affect the litigation... it would be news value for instance that at present the SMA has no legal standing in arbitration proceedings, even though it is paying the bills.

This last sentence referred to the fact the Authority had accepted, for the time being, awards set by the New South Wales Industrial Commission, at which the Authority had no legal standing and no ability to appeal against decisions affecting it—particularly decisions that granted increased wages or penalty payments.

Down but not yet out, the Authority tried again the following year. This time it lobbied the government to amend its Act, using a simple change of wording to give the Commonwealth Conciliation and Arbitration Commission the power to prevent and settle industrial disputes on Commonwealth projects, such as the Snowy. To do this, the term 'industrial dispute' was to be changed to 'a dispute as to industrial matters' and would apply even when a dispute did not extend beyond the limits of any one state. The Authority then filed a log of claims to bring the Snowy workforce under the jurisdiction of the Commonwealth Arbitration and Conciliation Commission.

But the case was unexpectedly put on ice by the sudden appearance on the scene of the South Australian government. It, too, had something at stake and had decided to put in its formal bid for a share of the water harnessed by the project.

This latest delay, and the Authority's dogged persistence in seeking a federal award, was the last straw for the Australian Workers Union. It had had enough of the political gymnastics and announced it would mount an all-out constitutional challenge to the power and functions of the Snowy Mountains Hydro-Electric Authority. (Charlie Oliver later stated he did not believe the action ran the risk of having the project shut down if the court ruled it unconstitutional. He felt some other way, such as channelling funds through New South Wales, would be devised to keep the project going.)

The Authority, on the advice of its own counsels—the Attorney-General, the Solicitor-General and the Crown Solicitor—withdrew its log of claims and surrendered. It agreed, finally, for industrial relations matters to remain under the New South Wales Industrial Commission, which is where the Australian Workers Union wanted them to be all along.

The Authority's backdown, and the union's subsequent withdrawal of its threat of High Court proceedings, settled once and for all the question of the Scheme's constitutional validity. After eight years, and about twenty drafts and redrafts of a formal agreement, the states eventually ratified the Scheme in their own parliaments.

The final agreement was loaded with a complex array of provisions and qualifications, but broadly it gave New South Wales and Victoria a guarantee that the charges for electricity from the Snowy Scheme would be based on the cost of production and would not exceed those they would otherwise have incurred in producing it themselves by other than nuclear power. If the Authority's production costs were above this amount, the Commonwealth would have to subsidise the difference. The federal government's acceptance of this, sixteen years before the Scheme was finished, was a considerable vote of confidence in the Authority's ability to estimate and control its costs.

Under the agreement, the Commonwealth would be solely responsible for financing the successive stages of construction. New South Wales and Victoria would contribute by paying for the electricity, as it became available, in yearly payments. For the two states it meant they would be able to meet the mounting demand for electricity without having to tie up millions of pounds of investment in additional plant. This finance aspect was very significant. By the time the state Acts ratifying the agreement had come into operation in January 1959 the Commonwealth's advances to the Snowy Mountains Hydro-Electric Authority totalled more than 200 million pounds.

The agreement detailed that the Authority would be responsible for the construction of the Scheme but that operation and maintenance of the completed power stations would be controlled by a Snowy Mountains Council made up of representatives of the federal, New South Wales and Victorian governments and the commissioner of the Snowy Mountains Authority. Victoria accepted sole responsibility for any measures needed to address the consequences in that state of the diversion of the headwaters of the Snowy River into the Murray River.

Diversion of a large proportion of the Snowy River was naturally going to reduce the flow through the river's normal channel into Victoria. Landholders on the lower flats along this river wanted protection from any adverse effects of the reduced flow. Murray landholders, too, were worried by the prospect that the extra water pouring down the Murray might at times flood their properties. South Australia was given a guarantee that during any specified drought period it would have a right to share all the water in the Murray River that was available for distribution, including the quantities diverted to it from the Snowy Scheme.

During the long period of stalemate, when the Authority was preoccupied by its shaky legal standing and its quest for a federal industrial award, it was fortunate that others had their feet firmly on the ground and remained in control of the day-to-day issues. It was during this period that Judge Taylor was most active, progressively modifying the various state awards to produce a de facto 'Snowy award'.

But the Authority was far from grateful, viewing the judge's activities with increasing antagonism. It felt he was too generous with the workers in their claims for special allowances for the type of work they were doing and the conditions under which they worked. It was also unhappy that the contractors had kept out of the issue completely, either actively or passively supporting the unions. By the end of 1956, it was estimated that Taylor's award modifications had already added about a million pounds to the cost of the work.

Yet when, at last, in 1958, after the states' ratification of the Scheme cleared the way for the Authority to reapply for a federal award, it held back from doing so. It looked a little more deeply at its industrial record under Taylor's jurisdiction and saw considerable merit. There had been many disputes but they had not caused work stoppages. The judge might well be costing them money in the immediate term but in the long run he was clearly establishing a record of industrial peace which would far outweigh, even in cost terms, the short-term concessions the Authority had to make.

Hudson was the only person who took a little longer to become convinced of Taylor's value. He once grumbled to the judge: 'You cost me a hundred thousand pounds every time I see you.' In the Authority's 1959 annual report Hudson wrote that a twenty-four per cent cost increase between 1954 and 1959 had been mostly due to increasing wages and that the wage increases on the Scheme were far in excess of those in New South Wales generally: 'The majority of these increases result from determinations by the New South Wales Industrial Commission [Judge Taylor] and relate specifically to the workforce on the Snowy.'

The judge took umbrage at the remarks and wrote to Hudson, expressing disappointment with the commissioner's attitude. Part of the letter read:

> Most of the workers are on over-award rates owing to the need for three shifts a day and all-weather work in generally poor conditions.
>
> If the statement by the Authority in the discussions at Eucumbene (about tunnellers' rates) that, 'this work will be done at award rates' had been permitted to be executed, Eucumbene and Tumut Pond would still be under construction. [Both had been finished for more than a year at the time of the letter.]
>
> Can any of the critics tell me where else do all workers, including carpenters, work shifts over a six-day week; where else do unions and men accept unreservedly the decision of the industrial authority?
>
> Recently at Sue City, a union official called a stop work meeting of metal tradesmen and I directed the cancellation of the meeting. The union and the men accepted my decision without question—a condition of affairs which made me feel very proud of them. Where else would it happen?

The judge, through his close association with Charlie Oliver, had 'educated' the unions about the ineffectiveness of strikes. Because of his presence on site, Taylor demonstrated that disputes could be settled fairly without the need to stop work and lose wages. He succeeded in getting the workers never to take action on their own account; always to work through the union which was in a position to represent them while they continued with their work.

Taylor let the workers know that he understood them. He believed that to get the best work out of men, they had to be well paid, well fed and well housed and have access to the amenities of life. He used his influence to gain these things for the men working on the Snowy, particularly as they had started the project in most horrendous conditions. Yet he demanded maximum performance and refused to support any worker sacked for breaching what he considered the work ethic.

He understood that the nature of the work—the spartan barracks, the snow, the dangerous roads and tunnelling—required a new set of industrial rules. He considered the Snowy project to be a unique situation in which the cooperation of all parties was essential and also in which work stoppages could not be tolerated.

In one of his decisions he wrote:

The conditions met here by this group of workmen are extraordinary. The work is consistently arduous and has to be performed with great care and indeed with much more than ordinary skill.

The safety of other workers to a very large extent depends on the way these men carry out their duties.

The principle I have followed throughout the Snowy area has been to accord a special rate where the conditions under which particular work is performed have special features to which the normal rates are not adequately applicable.

It is remarkable but nevertheless true, that in the Snowy area, there is no demarcation, or men refusing to do jobs which are not strictly, in their view, in the lines of their award, or classification; or indeed of men sitting around when work is to be done merely because it is not 'their work'.

This has resulted in a team spirit throughout the workforce, which, as far as I know, has no parallel anywhere.

When, in 1966, Taylor was reaching compulsory retirement age the idea developed that he be kept at work with the project as an independent arbitrator. But, as any decisions he made in this capacity would have no weight in law, there needed to be an agreement between the Authority, contractors and the unions that such decisions would be accepted and honoured. The Authority, eight contractors and eleven unions duly signed such an agreement. Under its terms, any party not willing to comply with one of Taylor's decisions could appeal to the New South Wales Industrial Commission. No one ever did.

As the states were finally agreeing to give their support to the Scheme, tenders were being called for the second group of major contracts for the lower Tumut development. This comprised the Tantangara and Tooma dams, the Tumut-2 underground power station and the Murrumbidgee–Eucumbene and Tooma–Tumut tunnels.

Tantangara Dam would impound the headwaters of the Murrumbidgee River for diversion through the Murrumbidgee–Eucumbene tunnel to Lake Eucumbene, the Scheme's central storage. Tooma Dam would create a reservoir on the Tooma River, a tributary to the upper Murray, for diversion through the Tooma–Tumut tunnel to Tumut Pond Dam, increasing the storage there for the Tumut-1 and Tumut-2 underground power stations.

To this stage, the American consortium had been granted almost all the major contracts. This, and the involvement of the United States Bureau of Reclamation, encouraged the Americans to regard the Snowy as a closed shop. They had produced good results and were popular, particularly with the commissioner. Furthermore, there appeared to be no viable competitors.

When tenders for the second major group of contracts closed, the Kaiser consortium quote was so high that it priced itself beyond consideration, despite its popularity with the Authority. There was a more realistic quote from another big American company, Utah Construction; but the cheapest quote was from a little-known Australian engineering firm, Thiess Brothers.

The Authority was unsure just what to do. Its heart was with the Kaiser group; the economics pointed to Thiess Brothers, but there was a big question mark over the Australian firm's ability to take on a project of this size. Then there was the Utah quote, which appeared to offer both capability and a reasonable price.

Before making a final decision, the Authority investigated the performance of Thiess Brothers on some of its past contracts. While they were comparatively small projects, the Authority was impressed by the high standards set by the company. It decided to take a risk. The Tooma Dam and Tooma–Tumut tunnel contracts were awarded to Thiess Brothers and the Tantangara Dam and Murrumbidgee–Eucumbene tunnel went to Utah.

That left Tumut-2 power station. There was strong support within the Authority for the retention of the Kaiser group, despite its quote. 'We were in an awkward situation,' recalled Ken Andrews. 'We wanted to retain Kaiser if we could because of their good performance on the first group of Tumut contracts; so we decided to amend the design to give them an excuse to re-tender with a reduced price.'

Consequently, the Tumut-2 Dam and Tumut-2 power station went to the consortium of Kaiser-Perini–Morrison–Raymond. Walsh Construction, which had been part of the first Kaiser consortium, had decided to drop out.

Thiess Brothers was the first Australian contractor to join the Scheme. The appointment was a gamble and even Sir Leslie Thiess later admitted to having been surprised: 'We had a bit of a cheek to go down there and take on something like that; but we got away with it,' he said in an interview some years after the Scheme was completed. The company had no experience, for example, in tunnelling and was forced to poach skilled workers from other contractors.

Thiess Brothers was started shortly before World War II by Leslie Thiess and his four younger brothers in the small Queensland town of Drayton on the Darling Downs. They took on any engineering work in the district, from providing power for threshers at harvest time to supplying building materials. After the war they reputedly 'made a killing' with scrap metal in New Guinea and by the 1950s had grown large enough to contemplate pursuing a Snowy contract.

As it turned out they performed well on the Tooma contract and the Authority had no hesitation about using them again. By the end of construction in 1974 Thiess Brothers had built a quarter of the entire Snowy Scheme.

TENTATIVE BEGINNINGS

Tunnellers excavating the site for the Tumut-2 underground power station, June 1959.

CHAPTER SEVEN

Ivan Kobal (left) with an Italian friend at Goodradigbee River Camp.

BUSH COOK

It was May 1950 when the *General Sturgess*, with Ivan Kobal aboard, docked in Melbourne. Ivan and the other refugees on board had left a Mediterranean summer and arrived in a drizzly Melbourne winter, but the climatic change merely heightened the sense of being in a new world.

Five traumatic years had passed since Ivan's long walk to Slovenia from school in Italy, but he now sensed he had found the refuge he had been looking for. He had felt it the moment the ship had put mainland Europe behind the horizon, so he had read and reread a dog-eared John Steinbeck novel to try and gain a better grasp of English.

From Melbourne the workers were taken to a camp at Bathurst in New South Wales. For most, camps of one kind or another had become so familiar that to go to another one seemed quite normal. But the Bathurst camp proved much more comfortable than any other camp they had experienced. 'There was so much space. We were put into cottages, each with its own little red mail box,' Ivan recalled. 'We didn't meet many Australians, but we were free to explore the countryside. "Rabbit Gully", "Kangaroo Creek" were the first English words many of us spoke as we explored and learned place names.'

On average, the migrants stayed in camp for three weeks while they waited to be placed in work. Most went to construction camps in the Snowy Mountains. Ivan, however, was sent to work in an iron foundry at Alexandria, a suburb of Sydney. A few months later he was transferred to a wire factory at Chiswick, where he was put to work making barbed wire. After leaving behind a world strewn with and divided by barbed wire, he could hardly believe the irony of coming to a free land where one of his first jobs was making the detestable strands—even though it was for fencing in sheep rather than people.

At the wire factory Ivan befriended an Italian, Franco Palumbo, who told him about the Snowy Mountains Scheme. Franco had friends working on the Scheme and reckoned he and Ivan should try their luck. The barbed wire was depressing both of them.

However, under the terms of their contracts with the government they had to work two years in the jobs they had been assigned. As the months crept by Ivan forgot about the Snowy and, on leaving the wire factory, sought work as a brickies' labourer. Every morning at sun-up he left his boarding house to ride an old pushbike from building site to building site until he became such a regular visitor that he was finally given work with a team that was building a church. The labour was hard and, with his limited grasp of English, he was looked down upon by the Australian tradesmen; but the job fired in him an ambition to become a carpenter. After the destruction of the war, he wanted to build.

During this period Ivan set himself the task of learning twenty-five new English words a day. These he wrote into a notebook.

He enjoyed the building work, but by Christmas 1954 his friend, Franco, had finally convinced him that they both should go to the Snowy. Ivan used part of his savings to buy a set of carpenter's tools. He had heard the Snowy was mainly interested in tradesmen.

Laden with his bicycle, new tools, two blankets and two suitcases, Ivan and Franco made the long, slow train journey to Cooma. The journey from Sydney took them through the misty Blue Mountains into the hinterland of New South Wales, through the fine wool region around Goulburn, past Canberra and into the vast rolling Monaro plains at the foot of the mountains.

The train arrived in Cooma at midday the following day. It was midsummer and hot. They left their possessions at the goods office and immediately found their way to the Authority's recruiting office. They came upon the queue long before reaching the office itself, and their hearts sank. The two friends joined the line of hopeful workers. After more than an hour of standing under the blazing sun and being harassed by legions of bush flies, the pair decided to collect their belongings and find a hotel room.

It was three days before Ivan finally gained an interview. He presented himself as a carpenter's assistant and proudly showed his gleaming new tools. The man behind the desk stared back without expression and asked to see his trade papers. Ivan had none. The Authority man laughed drily in exasperation and waved him away. As Ivan turned forlornly away the man called out after him, 'Wait up matey. I don't suppose you can cook, can you?' Ivan hated cooking; he had often gone hungry rather than cook. 'Sure, I cook good,' he said, beaming.

The employment officer took him at his word and sent him posthaste into the mountains. Food and cooks were a bone of contention among many of the migrant workers on the Authority's payroll. While the contractors had well-organised and expansive kitchens, the Authority had yet to come to terms with catering for the tastes of different ethnic groups. The Irish wanted big helpings of potatoes; the Italians wanted pasta; the Australians wanted big steaks. Any migrant who would try his hand at satisfying such diverse demands was more than welcome.

Ivan was despatched to a remote geologists' camp, the inhabitants of which had been threatening revolt unless the Authority supplied them with a decent cook. They got Ivan.

Refusing to be daunted by his lack of culinary expertise, Ivan served up a veritable mountain of spaghetti and onions for his first meal. The geologist in charge, an Englishman named John Newby, did not like it at all. But the European labourers and diggers thought it was tremendous, and outvoted the Englishman. Ivan had finally found a home.

The camp was at a place called Toolong Crossing, near the site of the proposed Tooma Dam. Ivan had to be up by six o'clock each morning, when the dew was heavy and the air chilled. He cooked breakfast for the workers in an old stockman's hut the team was using: 'It was cold and everybody had to wash in the freezing creek water but after they had all gone to wherever they were working, I had the camp and the bush to myself. My three companions were the wireless, the fire and the sun through the window.'

Ivan did everything with enthusiasm. He chopped wood, enjoying the vigorous activity in the cold air; he sang and whistled loudly and contentedly, knowing there was nobody to disturb; in his free time he explored the bush, fished in the river and wrote poetry.

After their evening meal of Ivan's spaghetti and onions the labourers for the geologists and surveyors, or diggers as they were called, sat near the fire, yarned and played cards by the light of a kerosene lamp. The diggers' task was to dig trenches into the topsoil to expose the first layer of rock as a preliminary to drilling.

After his own meal Ivan would walk through the dark bush, which was alive with the sounds of night insects and the distant melody of the tumbling creek, and say his

Ivan Kobal near Tooma Dam.

rosary. The confusion of the war and his years in the refugee camp were like distant memories. Here he was at peace; the world around him was at peace. Peace was no longer just an abstract notion. It had become real.

After a time even Ivan's staunchest supporters began to tire of his sole culinary offering; and he soon found himself working as a chainman and general hand while other workers occasionally took a turn at cooking. No matter how conscientious a cook may have been, it was difficult to maintain a varied menu in such a remote camp. Not surprisingly, there was great excitement whenever something as tasty as fresh rabbit presented itself as an alternative.

Winter was approaching when one morning Ivan joined a German geophysicist, Dr Bein, and his Czech chainman, Dusan Brydl, for a trip to Cooma in an old Austin truck. It was raining heavily and the three men had to squeeze onto the front seat.

Dr Bein was a shy, unassuming, middle-aged Berliner with only a minimal grasp of English. He had started on the Snowy as a cleaner in the Cooma office, until his considerable expertise was recognised. He was then promptly relieved of his broom and put in charge of a seismic survey unit. He relied heavily for social contact, though, on Dusan who spoke both fluent German and English. Dusan was full of confidence and, although the scientist was his boss, strangers meeting the pair for the first time often thought it was the other way around.

Dr Bein had determined to use the trip to espy and shoot a rabbit or two. He was fed up with slabs of cold lamb, beef stew and even Ivan's spaghetti and onions, on the occasions he visited Toolong Crossing. He sat hard against the passenger door with a rifle jammed between his knees, his eyes restlessly scanning the undergrowth. Suddenly, he let out an excited yell. Dr Bein had seen a rabbit. The other two peered through the rivulets on the glass, but could see nothing.

'Ah, it's gone...too far. It's no use,' said the German despondently.

'Don't worry. There will be more,' said Dusan confidently.

The journey had proceeded just another two hundred metres or so when Dr Bein again began yelling excitedly.

'Stop, stop! Back, back! Slow...shhh.'

The rabbit followed Dr Bein's commands to the letter. It stopped, raised its head and sat perfectly still while the scientist pulled the rifle from between his legs and pointed it through the truck window. It was the last moment in the animal's life. Dr Bein pushed open the cabin door and jumped excitedly from the truck to collect his prey. He carried it back, holding it high, a proud smile stretched across his face. He made a comical figure, a shaky body on skinny legs hidden by seemingly empty trousers. Dusan watched impassively, refusing to be impressed.

The rabbit was put on the back of the truck and the three men continued their slow journey, with Dr Bein now staring even harder into the dark undergrowth. They drove on, the mountain walls sometimes seeming to hide them even from the sky. As they crossed a creek in heavy rain, Dusan saw three wild ducks. The three men climbed from the cabin and trod quietly towards a fence rail, which Dr Bein used to support his arm while he took aim.

The first shot was unsuccessful and the heavy rain took the blame. Dr Bein took deliberate aim and fired again. The ducks remained unmoved. This was too much for Dusan, who demanded the rifle. He took a quick, impatient aim, and watched two ducks fly away while the third simply disappeared.

Dusan went off to recover the third duck, Dr Bein followed him, while Ivan watched with amused interest from the fence. The pair returned to the roadside, soaked and disappointed.

Ivan decided to eat the lunch he had prepared for the trip—rabbits legs, from one of his own hunting excursions. "Delicious," he told his two speechless travelling companions. Ivan thought it best not to tell his companions that in Cooma he would be getting more provisions—more spaghetti and onions.

After his first winter in the mountains, Ivan returned to Sydney where his old boss had offered him another job on a building site. But the bush had changed him and the following spring, clutching a letter from his Snowy boss, the English geologist, Ivan returned to the mountains.

He still wanted desperately to become a carpenter and tried to get a job in the dam models workshop. But it was impossible without formal qualifications. Finally, after exaggerating his experience and downplaying the fact that he was self-taught, he was granted permission to sit for the trades examination—and passed. Now relieved of the thankless role of bush cook, Ivan stayed in the mountains for another four years.

Food and meals were the focal point of all workers' day-to-day existence. Men of a dozen different languages and backgrounds could be united as one in their antagonism towards a particular cook, or in their despair at a popular cook's passing.

The workers at Tumut Pond camp felt particularly aggrieved one Christmas when the camp cook, known only as 'Old Percy', disappeared. Old Percy simply vanished. Christmas passed, with the men having to forego the special Christmas dinner they had all been looking forward to.

Nothing was heard of Old Percy until six weeks later when a group of canoeists, who had been making their way down the Tumut River, reported to police that at Christmas time they had found a body in the river near Tumut Pond. It was the first opportunity they had had to report the discovery. They told police they had tied the body by the leg to a tree above the waterline.

When word filtered back to the camp, the men retracted their unkind words about Old Percy's disappearance, and lamented the passing of a good cook. But, for some at least, the feelings of goodwill towards the old gent quickly dissipated when they were told they had to collect his remains.

The recovery team included the young Scotsman, Jock Wilson, who had transferred from Three Mile to Adaminaby; Ken Sharpe, a geologist; Harry Patchet, a medical officer, who liberated a large bottle of brandy from the canteen to help them all get through the task; and two policemen—Sergeant Cavanagh and a young constable—from Cooma.

The group had to struggle through thick scrub, descend a steep slope and cross the river to get to where the body had reportedly been secured. It was late afternoon by the time they found Old Percy and wrapped his decomposing remains in a tarpaulin. It was difficult to keep the now putrid body in one piece and, regardless of how fine a cook Old Percy had been, nobody in the recovery team felt they would ever be able to face food again.

They started to carry their burden up the steep hill to a point on the ridge, where the police had arranged to rendezvous with a vehicle. But the steep slope and dense scrub made it an awkward and frustrating task. By the time the party was halfway up the slope it had grown dark and nobody wanted to be the man under the bundle should someone stumble. So they left Old Percy once again by a tree, to be collected the next day.

At eight o'clock the following morning the men gathered at the top of the slope, this time with a pack horse borrowed from Major Clews's survey camp. But Old Percy was not where they thought they had left him. All morning they scoured the hillside. They did not find the body until mid-afternoon, by which time nobody was in a particularly happy mood. When Old Percy's remains were strapped across the back of the horse the men counted their blessings at finally being rid of the old cook.

The horse, however, was not at all happy about its role in the retrieval operation.

Geologist Ken Sharpe (left) and a chainman near the Tumut River, 1950.

Old Percy smelled pretty bad and the horse's nostrils flared in protest as it whinnied and tossed its head. Whether in relief that his job was done, or in frustration with the animal, one of the men slapped the horse's hindquarters. That was the last straw. The horse bolted, straight up the hill through the scrub, sprinkling little pieces of Old Percy behind it.

The men stood and watched the animal disappear. No one spoke. They dared not think of just where Old Percy would now end up or how long it would take to find him again. Their plans for burying the old man decently were going sadly awry. Too long on the back of the horse and there'd be nothing left to bury except a putrid piece of tarpaulin. The group climbed wearily back up the hill, following the horse's trail by the smell.

At the top of the hill, the men were greatly relieved to discover the animal standing by the vehicles. It too seemed to have tired of the exercise and gave no more trouble as Old Percy was unstrapped and put in the back of a utility. Sergeant Cavanagh volunteered to ride the horse back to the survey camp, where Jock Wilson met him with the ute and the pair returned to Cooma with what was left of the former Tumut Pond cook.

Old Percy was not forgotten in a hurry. The horse's back was soaked in the juices and bile from the body, all of which was infused into the fabric of the policeman's trousers and thence transmitted to the seat of the utility. Jock Wilson had to drive with the reminder of the cook's last journey for weeks until he was able to convince the Authority of the need to replace the vehicle.

In the mountains at Cabramurra, Australia's newest and highest town, Mario Pighin and Alfredo Fabbro, the two tradesmen from Udine in northern Italy, had become good friends. Both had been married just days before they sailed to Australia, leaving behind sweethearts they barely knew as wives. As they neared the end of their enforced two-year terms as single men, they had both worked vigorously to find ways of bringing out their wives, Angelina and Guiditta.

Mario found a housekeeper's job for Angelina with the area doctor, Dr Ina Berents, a Rumanian. Dr Berents, in fact, had been desperate for help. She had two young children and a third on the way. In addition to being a mother she was the only doctor in the region and the casualties from the construction sites were ceaseless.

Alfredo secured work for Guiditta as a cleaner in the staff mess. The two men arranged for their wives to travel to Australia together, and when the girls arrived at Sydney the couples joined up for a combined and belated honeymoon.

By this time, Mario and Alfredo had grown accustomed to the emptiness of rural Australia and the privations of the mountain workcamps, but for their two young wives the cultural shock and the strangeness of the environment hit hard. On the ride in the back of an open jeep from Cooma to Cabramurra, Guiditta winced with each new bruise and looked about with increasing trepidation. It was a cold, desolate landscape without people or villages. The jeep passed through Adaminaby, the only semblance of civilisation they had seen since leaving Cooma, and then continued its jolting, bruising journey to their new home—a crude townsite on a small plateau hacked from the side of a mountain. It was dark and snowing when, three hours after setting out, the quartet arrived at

Cabramurra; but the young women were happy, because they were at last with their husbands.

Neither woman spoke English and during their ten days together in Sydney, Mario had tirelessly coached his wife on what to say when he introduced her to Dr Berents. 'It is very important that you make a good impression,' he told her. 'When you are introduced you must say "Ow do you do" and then say "I am pleased to meet you".'

He tested her repeatedly: 'What are you going to say when you meet Dr Berents?'

'Ow do you do.'

'Good, good, that is good.'

Alfredo took Guiditta to the women's barracks and Mario took Angelina to Dr Berents' house, where the introductions were performed.

'How do you do?' Dr Berents inquired of her new housekeeper.

'I am pleased...to meet you,' said Angelina.

'Thank you. How was your trip?' Dr Berents continued politely.

'Ow do you do,' said Angelina, beaming.

It was 15 June, almost two years to the day since Mario had pedalled out at dawn in his best suit to meet the priest and marry Angelina before sailing to Australia.

Angelina started work the very next morning. As the men had found, there was no time to adjust gradually to the strange, new surroundings. The sudden realisation that she was also expected to cook threw Angelina into a panic. 'What do I cook?' she pleaded to Mario when he called by the next morning on his way to work. He shrugged his shoulders: 'Cook a bucket of spaghetti.' So Angelina cooked a bucket of spaghetti, plain with no sauce because there wasn't any. The Berents said it was very nice; so the family had spaghetti every night for the next month.

Though Mario and Alfredo had their wives with them, they were not allowed to actually live with them. They were 'wages' men and not entitled to married quarters. Angelina lived with Dr Berents and her family and Guiditta lived in a barrack with several German women. Angelina recalled:

> I was too busy in the day to be lonely but it was lonely at nights. Mario was so close but so far. I have never cried so much in my life as I did in those first months in Cabramurra. I couldn't speak English, there was no one to talk to and my mother and father were so far away. But when Mario asked if I wanted to go back home I said, 'No, just give me time.'

Six months after her arrival, Angelina became pregnant. With Dr Berents's three children and then her own, she was soon managing a sizeable nursery—not an easy job under the conditions, especially in winter. Because of their changed circumstances Dr Berents's husband Derek, who was an engineer, built Angelina and Mario a snow hut behind their house. It was small—just a single room measuring three metres by four metres—and had a bed, a stove, two chairs and a handbasin. When they got a cot they had to take out a cupboard to fit it in.

Despite the hardships, Angelina remembers the time as happy. The less people had, the happier they seemed to be:

Soon after we moved into the little hut, it was buried in a heavy fall of snow. We could open the door inwards but there was a wall of snow still blocking the doorway. When Mr Berents saw us emerge after we dug ourselves out he laughed and rushed to get his camera. It was freezing. Everything was wet and it could have been miserable; but instead it was something to laugh about.

For Guiditta and Alfredo the enforced separation when they were so close was hard to endure. For them it was even more difficult to be together because Guiditta worked in the staff mess, which was off-limits to wages men like Alfredo.

The situation was full of anguish for both. Guiditta had no one to explain away the strangeness of everything about her. The strong aroma of eucalyptus, which hung over everything, made her ill until she learned what it was and grew used to it. Guiditta was a trained nurse and the lack of stimulation in the hard cleaning work was mentally, as well as physically, draining. The work entailed cleaning the mess, polishing its vast wooden floors and cleaning all the staff cottages.

She also had to cope in the first few weeks with wearing borrowed clothes; the trunk containing her belongings had been sent to Como, in Western Australia, instead of Cooma.

At first the Authority refused point-blank to make an exception for Alfredo and allow him to go to the staff mess to visit his wife; but after many approaches and pleas it relented and told Alfredo he could visit Guiditta on the condition he wore a jacket and tie—and only on Sundays. The only times the couple really spent together were short periods after the evening meal, when Alfredo was allowed to visit Guiditta at the women's barracks. He had been threatened with instant dismissal should he fail to return to his own barrack.

Guiditta, for her part, was strictly forbidden to go anywhere near the men's barracks. Her room in the women's barracks was, like the men's, just big enough to accommodate a bed and small cupboard. There was a world of difference between this and the staff barracks which, apart from being more spacious, had self-contained kitchens, laundries and showers. The wages barracks could be up to a hundred metres or more from a laundry or shower block, which in winter meant trudging through snow and mud to and from the amenities.

All the women with Guiditta were married but similarly separated from their husbands, who were working on the Scheme.

There were three German ladies and me. We couldn't understand each other but we laughed a lot, especially when we were trying to learn English—'What you call this? What you call that?. . .' As we learned the names of things we'd tell the others. Whenever we didn't understand we would just say 'Yes' and smile.

The mess food was awful. The canteen served only lamb and mint sauce, or corned beef and cabbage. All the Italian men begged me to cook a good Italian meal for them, but it wasn't possible. We weren't allowed to cook in the barracks, though I kept a little pump-up kerosene stove and a saucepan under my bed for Alfredo and me.

When Alfredo visited at nights I would make some packet soup and coffee. If the mess meal had been very bad we would go into my room and cook some boiled eggs, soup and some coffee.

The couple then had to wait until no one was about to wash the dishes in the laundry. The little stove had to be kept locked in a suitcase under Guiditta's bed because the barracks were regularly inspected for tidiness and illicit cooking implements. Luckily for Guiditta and Alfredo, the camp kitchen was nearby to disguise the cooking aromas, and the stove was never discovered.

Every two or three days the women's rooms were searched to make sure their husbands had not stayed overnight. Inspections were sometimes sprung on the women when it was particularly cold, to ensure no one had smuggled in heaters, which were banned.

The barracks were heated by hot water pipes installed in the roof but, as hot air rises, they were totally useless for warming the frozen rooms below. The only use the women found for the pipes was drying clothes.

Eight months after Guiditta's arrival, Alfredo was promoted to leading hand. The couple's life and the lives of the many single young Italian workers improved immeasurably as a result. The promotion led to their being given a house with three bedrooms, a fireplace and a stove, both of which burned day and night. Because the house had a proper kitchen, the Italian workers were particularly happy: 'The boys would come by. "When you gonna cook us a good Italian meal, Guiditta?" But it wasn't that easy. There was nothing in the Cooma shops; no tomatoes, no oils, no vinegar, no pasta, tomato sauce, salami, garlic—nothing.'

The men overcame the difficulty by making sure that whenever anybody went to Sydney they took an empty suitcase in which to bring back all the supplies Guiditta needed to cook 'a good Italian meal': 'They would get the ingredients from Sydney, I would cook a big meal and we would have a party. It was the same if any of the boys got sick and were confined to their barracks. I would cook them a good home meal, an Italian meal, and take it to them to lift their spirits.'

Once settled in the house, Guiditta and Alfredo felt secure enough to start a family, and in 1956 their son, Robert, was born. The isolated work camp was not the ideal place to have a baby but Guiditta didn't even contemplate the possible problems:

> When I became pregnant I was too happy to be worried about anything. We were young and we had a house and a garden. Ours was the only Italian house where the Italians could get together and really party, so it was always fun.
>
> Alfredo made a bocce pitch and on the weekends all the Italian men would come to play, often late into the night under lights. If it rained they'd dry the pitch with sacks. They were very keen.

When Guiditta had become accustomed to the pungent odour of the eucalypts Alfredo sometimes took her camping and fishing.

Their lives were made complete by the monthly visits of an Italian priest, Father Boasso from Canberra, who would celebrate Mass in the picture theatre. Father Boasso would also bring an Italian film, which he screened after Mass. Then everybody walked to Guiditta and Alfredo's house for home-made pizza: 'All the Italians used to come for Mass, though I sometimes think it might have been more for the film and the chance to escape the mess food with my pizza.'

A christening at Cabramurra, 1955. From left, Carlo Del Dorso, Santo Del Santo, Mario Pighin holding Harold Berents, Guiditta Fabbro holding Nadia Pighin, Dr Ina Berents and son Derek, Mario Signarini and Alfredo Fabbro.

CHAPTER EIGHT

Drillers aboard a three-deck drilling jumbo in the Tumut-2 tailrace tunnel.

TUNNELS OF BLOOD

On the very day Derek Berents had laughingly taken photographs of Angelina and Mario Pighin digging their way out of their little snow hut, the shrill ring of the telephone had summoned his wife into the snow-covered bushland to yet another tunnel accident at Tumut Pond.

Dr Berents knew the landscape well by now but rarely had time to admire the alpine beauty as she skied cross-country to render what emergency medical aid she could.

As the stout figure of the fair-haired woman sped through the snow gums, her ski stocks flaying the fresh snow, a New Zealander, Jack Roden known by his workmates as 'Kiwi Jack', was lying in a pool of blood and vomit beside the railway tracks inside the tunnel that was being blasted to link Tumut Pond reservoir with the Eucumbene Reservoir.

The blood was Kiwi Jack's; the vomit was that of some of his workmates, who had rushed to his crumpled figure after he had stepped in front of a tunnel locomotive. One of the first to reach him was an Australian powder monkey, Alan 'Happy' Hetherington, who after finding a pulse had gently removed the injured man's helmet. With the helmet, however, came the top of Kiwi Jack's scalp, exposing his brain. His stomach heaving, 'Happy' Hetherington gingerly replaced the helmet and sounded the alarm.

The injured man was removed from the tunnel on a stretcher jeep, which took him to the first-aid post to await Dr Berents and an ambulance. The ambulance had to come from Cabramurra, slowly behind a snow plough, and then descend a narrow, treacherous track which dropped six hundred metres from the ridge to the camp. It was one of the reasons Dr Berents preferred to ski. To her, it was quicker and usually a lot safer to cut across country on skis than try to get through by road.

As she pushed past the exposed, leafy tops of buried snow gums, the doctor, however, had more on her mind than just the injured man, who from past experience she knew would quite likely be dead by the time she arrived. Before getting the urgent summons from the tunnel, she had already been about to leave for Tumut Pond where the pregnant wife of one of the workers had gone into labour.

So she had two cases, and the weather seemed to be worsening. It was going to be a difficult day; just how difficult she did not realise until she arrived. To her horror, the injured tunneller and the pregnant woman were husband and wife.

She tended the woman and did her best for the man, issuing instructions that on no account was the woman to know the identity of her fellow passenger in the ambulance. When the vehicle finally arrived, she screened the two patients from each other.

The trauma, however, was just beginning. Half-way up the steep climb out of the gorge, the ambulance skidded and slipped onto its side in a snowdrift. Doctor, patients and equipment finished up tumbled against each other inside. Dr Berents disentangled herself and, with the help of the driver, lifted the patients from the vehicle and laid them on stretchers in the snow. If ever a doctor had reason to pray for miracles, it was Dr Berents. Fortunately the husband was unconscious, making it easier for her to devote her attention to the woman, who still did not realise the injured man with the shocking head injuries was her husband. It was unlikely she would have recognised him anyway behind his mask of bloodied bandages.

Leaving Dr Berents huddled in the snow with her charges, the driver staggered off down the track to fetch a bulldozer to pull the ambulance free and tow it to the top of the mountain. It was some hours before the ambulance finally wheeled into the casualty entrance at Cooma Hospital. Miraculously the man was still alive and the woman had not given birth. Husband and wife were rushed into separate parts of the hospital for their respective emergency treatments.

Despite their ordeal, mother and child were the next day reported healthy and well. Even more amazingly, Kiwi Jack was back at his old job in the tunnel just seven weeks later, the top of his head stitched firmly back in place. He told his workmates that he had heard them talking about the top of his head and his exposed brain but had been trying to tell them that his head felt okay; it was his back that was hurting.

Kiwi Jack was one of the lucky ones. Countless others were killed or crippled—victims of an overriding obsession with speed.

In 1958 Sir William Hudson told a national conference on industrial safety that an analysis of accidents on the Snowy showed that the use of modern, fast-moving plant and the increased tempo of work on large, modern construction sites were the greatest sources of danger. He believed work safety education had not yet caught up with the technological and mechanical advances now employed on construction sites. He said these were new circumstances to which workers had yet to adjust.

Sir William told the conference that more than half the fatalities on the Snowy were attributable in one way or another to fast-moving plant. 'But it would be wrong to slow down the rate of work,' he said. He argued that reduced accident rates lay in a closer observance of safety procedures.

In 1959 the Authority set up the Snowy Mountains Safety and Rehabilitation Council to oversee industrial safety on the Scheme. But the practicalities defeated the good intentions. The contractors had their eyes fixed firmly on the bonuses the Authority was offering for finishing projects ahead of schedule. In striving to achieve this, they in turn offered flow-on bonuses to the workers, giving them tacit encouragement to take short cuts and ignore safety procedures.

Shifts were pitched competitively against other shifts. Daily and weekly progress became the talking point in the messes and even in the comparative remoteness of Cooma, where the local newspaper picked up the fever generated by attempts to set new world records for tunnelling progress. As a result, scant regard was paid to safety. There was no formal training in tunnelling, and the few industrial safety films in existence were in English.

The Authority had safety supervisors deployed at most construction sites but such men were in an impossible position, receiving no thanks from anybody for stopping work over a safety concern. The contractors disliked delays; the Authority preferred they did not happen; and even many of the workers saw stoppages in terms of lost money, even if it was for their own welfare.

Anything and everything was justified in terms of progress. This gave the contractors virtual carte blanche in their treatment of workers. The Americans, in particular, seemed to care little for the men, particularly the migrants, who in their eyes were merely overpaid 'wetbacks', or construction-site mercenaries, to whom the dangers were fair risk for good money.

This brutal and arrogant attitude may have cost many lives, but it was sanctioned by the Authority because it achieved results. The acceptable rate of fatalities on such projects in other parts of the world was three deaths for every mile of tunnel. Today, the Authority prides itself on its record of only one death for every mile of tunnel—but the fatality figure on which this statistic is based is open to serious question. Not recorded are the numerous road fatalities or the many disabling injuries sustained on the Scheme—enough to cause the Authority in 1963 to set up its own rehabilitation centre for injured workers.

At every construction site the strident, hustling voices of the American supervisors could be heard, often above the noise of machinery. 'Go, go, go!' they yelled as they strode square-jawed from work gang to work gang, like part-time actors enthusiastically performing their only line in a bad script.

As an example of their unbending attitudes, the Americans would order scaffolders on dam walls to use both hands on the job. 'Goddam, that's what the harness is for,' would be the response if someone dared complain that he might need one hand free to support himself.

One day the workers did complain about this requirement and stopped work over the issue. An AWU representative climbed high into the scaffolding against the vertical face of a dam to confront an American supervisor who was trying to enforce the rule. The union man told him the rule was ridiculous, and that men working hundreds of feet off the ground needed one hand free. The supervisor, a big bull-necked man, angrily retorted that it would slow the work. The union representative started to remonstrate but the American settled the argument by grasping his collar and thrusting the man over the edge of the formwork, threatening to drop him unless he instructed the men to return to work. The incident was witnessed by dozens of men but there were no repercussions for what was, in intent and purpose, a death threat. The men resumed work, putting their faith in the flimsy straps that held them to the scaffolding.

Often, however, the union was as unsympathetic as the contractor. Union secretary Charlie Oliver was as intolerant of strikes as was the Authority. He often argued that the men were fed and paid well and that was enough. Oliver did not particularly like the type of worker generally found on the Snowy. He made comparisons with the coal industry. Coalminers were loyal to an industry, but Snowy workers, particularly the tunnellers, he also regarded as no more than mercenaries.

Another common practice aimed at speeding progress in tunnelling was to ignore the supposedly mandatory waiting period, during which exhaust fans removed dust and fumes after blasting, before workers returned to the rockface to recommence drilling.

Because of the time bonuses, the workers jumped straight back on the locomotive and returned to the face, resuming work in zero visibility and also filling their lungs with dust and poisonous fumes. International health regulations disallowed work in an environment containing more than 250 parts per million of dust concentration. The tunnels, after blasting, measured more than 4000 parts per million.

Months after finishing work in the tunnels, men were often reported to be spitting black mucus from their lungs. Many later contracted lung diseases such as silicosis. A typical case was that of Con Dumitrascu, a 'displaced person' from Rumania, who at one time was a member of a tunnelling team which set a new world record.

Con contracted a lung ailment as a result of the tunnel environment and had to have a rib and one-third of a lung removed. While he was recuperating, a union representative visited him and told him the union could get 12 000 pounds compensation for him. In the late 1950s it was an enormous sum. But there was a catch. The union would only act if he agreed in turn to give half the amount to the union. Con decided to forego compensation.

With such uncertain union support, the contractors and the Authority wielded total power. The Americans were openly feared by many of the migrant workers. They had control over a man's life and livelihood.

On one occasion the Kaiser group was using a bulldozer to fill a gully in order to build a flatter road, Snowy Ridge Road, over the range from Cabramurra to Tumut Pond Dam. It had been raining heavily and the bulldozer operator was having trouble getting his machine up from the bottom of a loose and steep slope. After several failed attempts, he let the machine slip back to the bottom of the gully, where he abandoned it and climbed the hill on foot to discuss the problem with the foreman. The two men studied the ground and the foreman felt it would be best to bring in another bulldozer to pull out the stranded machine. But the driver felt such a recourse would reflect badly on his ability. So he scrambled back down the slope to try again, his footfalls starting dozens of small, gravelly avalanches. The foreman watched, his heart in his stomach, as the man wrestled with the giant machine, forcing it to claw its way up the steep climb. The machine was just metres from the top when it tipped backwards. There were no safety cabs on bulldozers in those days, and the driver was crushed to death on the first roll.

The bulldozer finished on its back at the bottom of the gully. The next day a crane winched the machine up and it was taken to the Kaiser workshops where an Italian mechanic, Alessandro Wialletton, was working. Shortly after the bulldozer was brought in, the company operations manager walked in: 'How much damage? How soon can you have it fixed?' Alessandro was still thinking about the man who had been killed. Something in the American's tone made him react: 'What do you mean? A man is killed and you worry about a bulldozer. What is going on?' he asked in broken English. The American rounded on him venomously: 'God damn you, that dozer cost us 12 000 pounds. The guy cost us nothing. So you just get that goddam machine working, buddy.'

He slammed the workshop door behind him, leaving the mechanic angry and frustrated: 'You never saw the bosses like him down in the tunnels, or trying to operate a machine high up on a cliff,' he later complained to friends.

At the time, however, Alessandro held his tongue. It had taken him three years to get a job on the Scheme in his chosen trade and he didn't want to throw it away. He turned his attention to the dents and minor buckles and shut his mind to the reddish-brown smears around the top of the seat.

When he arrived on the Scheme in 1953, after working out his two-year Government contract in Sydney, Alessandro's qualifications as a mechanic had been pointedly ignored: 'It was pick and shovel work, and at nights I cried. If my father, who hadn't wanted me to come to Australia, had seen me then he would have been ashamed. I was a trained motor mechanic but no one would listen. They just handed me a shovel. It was that or nothing, they said.'

His first job was on the construction of Cabramurra, after which he was transferred to Cooma. He left the Scheme when he was offered work as a mechanic on the construction of a dam at Muswellbrook:

> But it was totally disorganised. There wasn't even any accommodation for the workers. So I returned to Sydney and saw an advertisement by the Public Works Department, which was looking for mechanics. I joined the department and learned it had the job of building the dam at Eucumbene, so I asked if there was any work going there.
> The foreman was surprised: 'You mean you want to go there, down there in the bush?' He couldn't believe anybody actually wanted to go there. The department was offering all sorts of incentives to try to get workers to go to Adaminaby—free travel, a paid three-day break in Sydney every three months and special bonuses.

Alessandro went happily, glad to be back in the mountains. A year later the department was sacked by the Snowy Authority and Kaiser-Walsh-Perini-Raymond took over the contract. Alessandro moved across to join the contractor's workforce. He stayed with the company and later the same year started work as a maintenance mechanic in the Eucumbene–Tumut tunnel at Junction Shaft.

Junction Shaft was constructed about a third of the way along the line of the Eucumbene–Tumut tunnel to open up an extra two faces, in addition to those being worked from each side of the range at Eucumbene and Tumut Pond. Men and equipment, including locomotives and wagons, were taken down in a massive lift, raised and lowered the ninety-one metres from a giant gantry. The same lift also brought to the surface the tunnel mullock.

The tunnels were dark and wet, illuminated only by a string of electric light globes along one wall, until one neared the face. There it was like a crazed filmmaker's impression of hell. Men in glistening oilskins moved disjointedly like demons through the flashing blue strobes of welding torches. Beyond them, at the face, powerful arc lights threw a harsh glare over ant-like teams working on giant two- and three-deck work platforms, or jumbos as they were called.

Drillers clung desperately to the end of four-metre-long drill bits which penetrated with a harsh deafening rattle into the rockface. They wore no ear protection and their faces were raw from the stinging spray of their water-cooled tools. Moving among them in the confined space were locomotives which shunted wagons and collected rock spoil. A man only had to take one wrong step, or stumble, and he could be crushed and dismembered under the steel wheels.

The locomotives used in the trans-mountain tunnels were sixty-horsepower, battery-driven electric engines which pulled an average of seven wagons, each containing four and a half cubic metres of rock fill. This was standard throughout the Scheme except in the Murray-1 and Snowy-Geehi tunnels in which diesel locomotives were used.

To Alessandro and hundreds like him, the tunnels were worse than a nightmare because they could not wake up to escape:

> It was frightening, terrifying. The only sound was the ear-splitting noise caused by

Construction of the entrance to the Eucumbene–Tumut tunnel at Eucumbene Portal.

Opposite
Tunnellers being lowered down Junction Shaft in a mullock bucket.

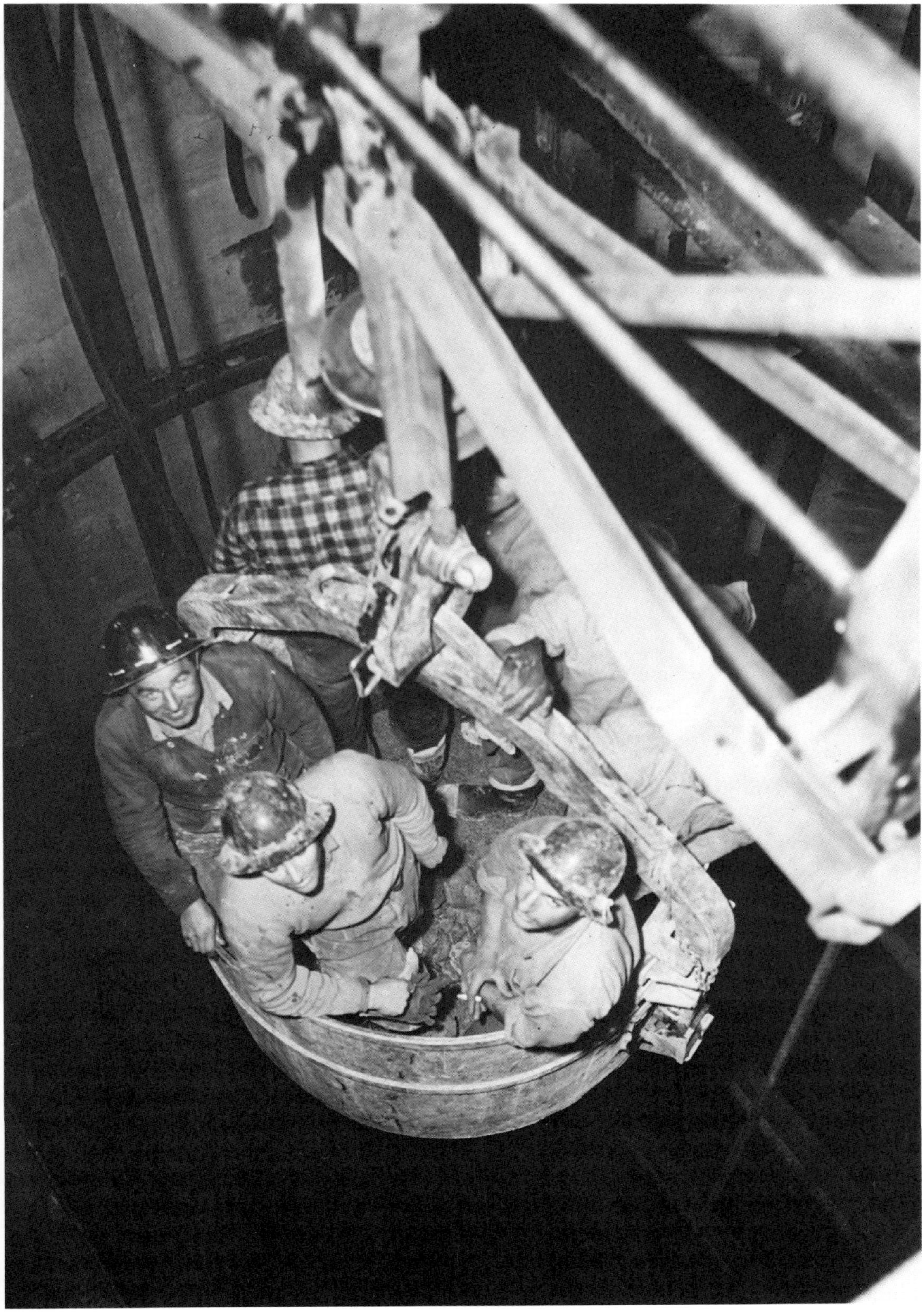

Drillers at the face of the Eucumbene–Tumut tunnel, November 1955.

Tunnellers at the face of the Eucumbene–Tumut tunnel, 1956.

drills and compressed air. The fumes made it hard to breathe and you could barely see your hand in front of your face in the vapour and dust which filled the chamber. Fresh air at the end of a shift was a joy in itself.

The floor was slippery rock, covered in a film of oil, grease and water. One day one of the loco guards jumped off to run ahead to switch the track. As his rubber boots hit the ground he slipped forwards and the loco's wheels cut off both arms. It happened in an instant and it was so typical.

Sometimes it didn't matter how careful you were. If a rock dropped from the roof just as you walked under, what could you do? If a compressed air hose burst just when you were using it, what could you do? You worked long hours, you were always tired, and when you were tired your concentration lapsed and that was it.

Colin Purcell, a tunnel inspector with the Authority, was almost the victim of a locomotive accident in an incident that, in retrospect, is rich in humour.

Purcell was sitting inside a portable lavatory, not taking too much notice of the low rumble and vibration that signalled the approach of a locomotive. The lavatory was tucked inside extra space blasted from the tunnel wall to make a work area, and was at a comparatively safe distance from the tracks. But what Purcell could not see in his contemplative confinement was a long piece of broken drill bit that protruded from a large chunk of rock on one of the wagons. It hit the lavatory with great force, tumbling it end over end for about twenty metres. Purcell was lucky. It could very easily have killed him. Instead, he emerged covered from head to toe in excrement: 'I staggered to my feet, shocked, and approached the locomotive, which had stopped. I tried to climb aboard but the sight and smell must have been pretty bad because the driver started screaming, 'No, no' and gestured frantically for me to go and climb up on to one of the wagons.'

Alessandro almost died one day simply because someone forgot to turn on the exhaust fan in the tunnel. He and two others were sent to the tunnel face one Sunday to make repairs on a drilling jumbo:

> We started to work and were feeling tired, but it was normal to feel tired. Everybody was always tired. Sometimes when there were problems you worked twenty-six or twenty-eight hours before getting any sleep. So we drifted into unconsciousness without even knowing anything was wrong. It was Sunday; no one else was about. But by chance a supervisor had come into the tunnel to see me. On his way in he noticed the fans were off and turned them on. When he reached us we were drowsy and just coming round. If he had not switched on the fans, he would have found only our bodies.
>
> Such things happened and you were scared. One time an air hose burst and started scything with enough force to cut a man in two. It had happened before. One fellow panicked—he had seen his mate 'shot' through the chest and killed by a sharp stone fired from the ground by the force of a broken air hose. So he jumped head first into the bucket of the mucker to escape, but knocked the controls on the way. The bucket lifted up against the roof and cut off both legs.
>
> It was the most terrible thing I have ever seen. After you saw something like this happen, you had days when you were so scared you could hardly move. I was frightened every time I went into the tunnel. In the back of your mind you always wondered if you would see sunlight again. I used to thank God every night that I was still alive. I had seen so many friends killed.
>
> Often you didn't know what had happened. You just saw the ambulance drive away and all the men leaving work. You knew someone had been killed and so you also stopped work.

It was a practice for all work to cease for a day when a man was killed. Work also stopped to allow workmates to attend the funeral. If an Irishman was killed the wake would sometimes last for days. It took time to get over the shock of seeing a mate cut in two or blown to pieces; but to the contractors and the Authority, accidents were time and money lost, not personal tragedies.

People today ask why men stayed in the tunnels. The underlying reason was the money. With bonuses a tunneller could earn up to fifty pounds a week—double the average wage. Tunnellers were urged on by footage bonuses on top of their hourly rates. When the going was good a man could make more from footage bonuses than his ordinary pay.

Some men were earning so much in the tunnels that money almost lost its value for them. They spent all they earned on women and alcohol. In Kings Cross, Sydney, the

Tunnellers awaiting a firing at the face of the Eucumbene–
Tumut tunnel, November 1955.

ten-pound note became known as the Snowy Dollar.

Four Portuguese miners at Tantangara one day chipped in and bought a new Holden. It was delivered to the camp, with just the delivery miles on the clock, and on their first day off they piled inside and, in a cloud of dust and bricks, sped off to Cooma for the day. They had gone just a few kilometres when the driver lost control on a bend and the car overturned down an embankment. Though the car was a write-off, the four men clambered unhurt from the wreckage and walked back to camp. To the astonishment of onlookers they then telephoned for a taxi from Cooma, one hundred kilometres away, to collect them. They were more concerned about losing their day off than the money they had just lost by wrecking the car.

With such expensive habits many a man came to rely on big wages. Some were forced to stay in the tunnels, earning big money to feed gambling habits. Illegal gambling dens did a flourishing business in the mountains. Others were there because they had little choice. If they were migrant workers under a two-year government bond, they had to work where directed. Men who had been refugees and displaced persons had arrived in Australia with nothing. This work with its high wages was their last chance to build some sort of future.

For a long time, Alessandro, who today runs a garage in Cooma, could not bring himself to look at the memorial erected to those killed:

I didn't want to remember the fellows I had come to Australia with and who were dead so quickly. It hurt, still hurts, to remember. A friend was beside you one day and dead the next.

People look at the Scheme today and think, yes, it's good. But a lot of us look at it and remember our friends and the places where we lived and where our friends died—places that have now been given back to the bush so that you can't even tell anybody was ever there.

The procedure for building a trans-mountain tunnel began with the surveyors, whose measurements and calculations were crucial for tunnels which had to meet up at a point deep under a mountain range. The work started on the surface with surveyors working in conjunction with diamond drillers, who determined the rock condition at the proposed tunnel depth.

Once the course of the tunnel was decided, the excavating began—from both sides of the mountain at once. It was at this stage that the surveyors really earned their keep. If the tunnels failed to meet, millions of pounds would have been wasted—and the public embarrassment would have been enormous.

On the Snowy the tunnels met every time, often to within a couple of centimetres. There was, however, a slight hiccup on the final breakthrough for the first trans-mountain tunnel, Eucumbene–Tumut, on 26 August 1957. The role of the 22-kilometre tunnel was to link the main Lake Eucumbene storage and the Tumut Pond Dam, which was to be the headwater for the Tumut-1 power station. With a diameter of seven metres, it had a carrying capacity when finished of 113 million cubic metres of water a second.

The Eucumbene–Tumut tunnel was dug from four faces—from either side of the mountain range at Eucumbene Portal and Tumut Pond, and from two internal faces heading in both directions from Junction Shaft. When the time came to make the breakthrough, there remained only two faces, one each side of the final three metres of rock that separated the Junction Shaft and Eucumbene tunnel sections. At this point the Junction Shaft tunnellers finished, leaving the Eucumbene crew to make the breakthrough.

The Junction Shaft tunnellers, pleased with a job well done, put their feet up in the Happy Jacks wet canteen where they discussed the culmination of five years of tortuous labour. Happy Jacks was the township built for the tunnelling from Junction Shaft, and also for the Happy Jacks Dam built later to divert the Tumut and Happy Jacks Rivers down the shaft into the Eucumbene–Tumut tunnel. It was named after Welsh gold miners who were there during the Kiandra goldrush days.

There was always great excitement when the final wall was to be broken and men who had been blasting their way into a mountain suddenly stood face-to-face with others who had been doing the same from the opposite side of the range. They were joyous both at their success and their survival.

The breakthrough of this, the first tunnel, was a major event, and one to be marked with pomp and ceremony. The workers were imbued with a sense of importance and pride in their achievement. After the final detonation, and when the dust had been sucked out and the roof cleared of loose boulders, the premiers of New South Wales and Victoria—

TUNNELS OF BLOOD

Visitors and miners at Eucumbene–Tumut tunnel breakthrough.

Premier of NSW, Mr Cahill, and Premier of Victoria, Mr Bolte, firing the last face of the Eucumbene–Tumut tunnel, 26 August 1957.

The Premier of Victoria, Mr Bolte, shaking hands with the Premier of NSW, Mr Cahill, at the Eucumbene-Tumut tunnel official breakthrough ceremony, 26 August 1957.

Surveyors marking the face of the Eucumbene-Tumut tunnel.

Joe Cahill and Henry Bolte—were to enter from either end, meet and shake hands.

Buoyed by the sense of occasion, the men celebrated. Inspired further by several hours of steady drinking, they boisterously took up a suggestion by one of their number to make one last nostalgic visit to the face and to listen to how the fellows on the other side were progressing. In high spirits, they rode the lift down Junction Shaft back into the tunnel—the place that had killed eight of their workmates and whose dark confines every man had been privately relieved to get out of for the final time. They commandeered a locomotive and rode to the face.

There they listened to the rattle of drilling from the other side—then looked at each other, aghast. The noise seemed to be coming at an angle away from the line of the face. After much pressing of ears against the cold rock they convinced each other that the two tunnels were not going to meet as exactly as they had believed. They reckoned the drillers on the other side were going to break through about half a metre to one side. They clambered aboard the locomotive, sped back along the tunnel, took the lift to the surface and excitedly leapt into an assortment of vehicles to race to Eucumbene

The Prime Minister Sir Robert Menzies posing for East German refugee Dieter Amelung outside the Lodge, Canberra, late 1955. The low hill to the right of the picture is now the site of the new Parliament House.

A semblance of order and civic pride—Cabramurra, winter 1959.

Opposite
A Norwegian worker showing the locals another use for snow.

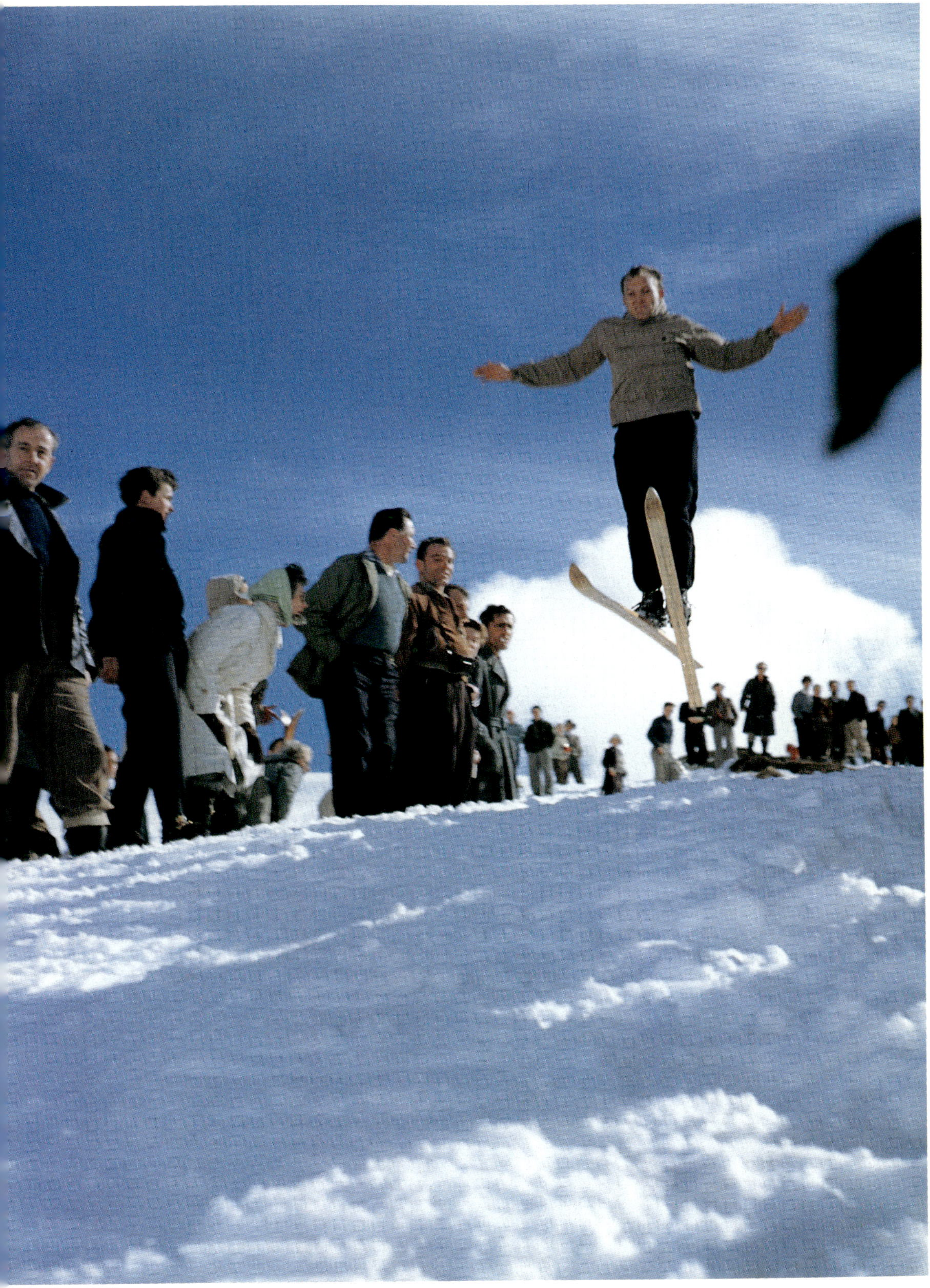

Karl Heinz Birnbaum (left) and Dieter Amelung manning the Cabramurra diesel power station in 1954.

A frosty morning outside the Kiandra chalet which denied shelter to Jock Wilson and his troop of foreign workers. The chalet and the rest of the township have since been bulldozed away.

Portal with the news.

Without bothering to enlighten any of the engineers or surveyors, they went direct to the tunnel supervisor at the face. The drillers there took some persuading but finally agreed to swing their drills across to change the final direction by the amount it was reckoned the tunnel was out.

The following day the dignitaries were assembled and the last shots fired. When the dust had cleared the VIPs applauded heartily. To them, the tunnels had met perfectly. But the drillers stared incredulously at what they recognised as a near miss. Had the Junction Shaft crew stayed in the canteen and not returned for a final curious look at their handiwork, however, the two tunnels, driven in from opposite sides of the mountain, would have met on a matchbox.

The responsibility for ensuring the Snowy tunnels stayed on course rested with five German surveyors: Wally Wassermann, Henry Kirsch, Fred Georg, Henry Werner and Bill Mueller. They were backed up by two Australians, George Bennett and Frank Johnston, and a Dutchman, Peter Ricardus. George Bennett later became professor of surveying at the University of New South Wales and Peter Ricardus became a senior lecturer there.

Six surveyors—George Bennett; Fred Georg; Henry Kirsch; Wally Wassermann; an Englishman, Peter Williams; and another Australian, Ian Foxall—were credited with devising a simple, failsafe procedure for keeping the tunnels on line.

The direction of the tunnels was based on two measurements—horizontal alignment and vertical level. Both had to be exact. The tunnels could be perfectly aligned but if the levels fell out, one could pass over the top of the other. Tunnels that did not run in a straight line were even more difficult to align.

The six surveyors decided that vertical and horizontal control would be carried out by three different teams who would make what they termed third-order, second-order and first-order checks. The third-order check was completed as the tunnellers were drilling into the face for blasting. Every third or fourth day the third-order surveyors would give the alignment measurements to a face painter who would actually paint the horseshoe outline of the tunnel on the rockface to be drilled and blasted. These third-order surveyors checked their progress in stages of between 150 and 200 metres. One of the surveyors involved in this work was Kon Martynow, the young refugee who had worked in the mountains on the early surveys with Major Clews.

Following the third-order surveys, the Authority's chief engineer would periodically ask the contractor to give the second-order survey team access to the tunnel for half a day. It needed to be a formal request because the contractor, geared to work twenty-four hours a day, seven days a week, three shifts a day, did not like being interrupted.

The second-order team, containing no one who had worked on any part of the third-order work, would enter the tunnel, set up a theodolite in the middle of the rail track used by the mullock train and take precise measurements every 600 to 800 metres. The second-order team used equipment which could check vertical and horizontal control to an accuracy of half a millimetre. They fixed their measurements with cement markers embedded along the centre of the tunnel floor.

The first-order checks were even more precise. Twice a year, on Easter Sunday and

Christmas Day, this group would spend the whole day measuring along the markers left by the second-order surveyors, checking the accuracy over even greater distances. Because of the darkness and poor visibility they used light points to take their sightings. This system of three independent checks ensured that the alignment problems commonly experienced at the time in tunnel projects elsewhere in the world were successfully avoided.

Once the tunnel was started, rail tracks were laid—first for the big drilling jumbos, which worked at the face; and then for the locomotives. The tracks were laid by a group known as the 'bull gang'. The bull gang was also responsible for running water and fan lines to the face and hosing down the muck pile as soon as possible after blasting.

The jumbos were hinged in the middle so they could be swung back to allow the 'mucker' through to the rock spoil after blasting. The mucker was an electric-powered machine with large steel buckets; it worked on a conveyor. The front could be swung from side to side across the width of the tunnel. The buckets scooped up the rubble, tipping it into the muck cars to be hauled away by locomotive.

It was a continuous cycle from midnight Sunday to midnight Saturday—drill, pull the jumbo back, blast, bring in the mucker, then push the jumbo back for the next drill pattern. Sunday was used by maintenance teams to inspect and maintain the rockface equipment.

As a tunnel inspector, Colin Purcell's job was to make sure the contractor worked to the specifications—particularly the rock support specifications—set by the Authority's engineers. Sometimes the rock structure was sound enough to leave alone; sometimes it needed rock-bolting; and sometimes it needed a steel set—arched ribs backfilled with concrete. If reinforcing was considered necessary, an inspector often had to stand his ground against determined opposition.

'Reinforcing a tunnel cost time, so if the contractor thought he could get away with not lining, he would,' said Purcell. 'Their supervisors were often just promoted miners. There were no Rhodes Scholars among them, so there was a lot of variation in how well they thought through a particular situation.'

As the Authority's scrutineers at the sharp end of the operation, the tunnel inspectors lived with the dangers of the location. 'The rules were bent, no denying that, so you had to be careful,' said Purcell:

Generally, you worked in teams so you could watch each other's backs. It was a harsh environment at the face. Newcomers needed mates to watch out for them until they became acclimatised to the conditions.

With up to thirteen drilling machines working on a jumbo, visibility would be cut to just a couple of feet after fifteen minutes through fog from the water-cooled drills, dust and exhaust fumes—not a very pleasant breathing atmosphere either.

It worsened what was already poor visibility, remembering the only illumination in the first place were lamps. Changing thirteen-foot drill bits in two feet of visibility meant you needed to know what you were doing.

Drillers worked in pairs, one man on the drill and the other at the face holding it against the rock until it was deep enough to stay there. The drillers charged their own holes, while a powder monkey—not necessarily someone with an explosives certificate—made up the primers. When loading the charge into new holes, the work was illuminated by the light of the locomotive. It was too dangerous at that moment to use electricity.

A pattern of charges was fired with half-second delays. It began with the centre of the face, then spread out to the perimeter, and finished with a bottom 'lifter' charge. Blasting the centre first created an internal face on which the outside charges could implode.

In good rock the tunnellers would fire a full round over the entire face to the maximum drill depth. One such round would advance the tunnel about three and a half metres. With two rounds a shift, and three shifts a day, a tunnel on a good day could be extended by twenty-two metres. If the tunnellers were in bad ground which was going to need reinforcing, they drilled a minimum number of holes and set small charges with centres of between half a metre and a metre.

Before the Snowy, progress of twenty to twenty-five metres a week was considered adequate by world standards. But as the Kaiser group progressed with the Eucumbene–Tumut tunnel it announced achievements of 111 metres, and then a staggering 149 metres a week. Then, in September 1959, the Australian contractor, Thiess Brothers, announced a new world record of 161 metres in the Tooma–Tumut tunnel.

The record stood for just a week before contractors on a Scottish project told the world they had achieved 171 metres in a seven-day period. The Snowy tunnellers were whipped into a near frenzy, and a week later Utah extended the Murrumbidgee–Eucumbene tunnel by 180 metres in seven days.

While the majority of fatalities involved locomotives and premature explosions—most of them caused by drilling into holes that contained unexploded charges from the previous firing—rock falls, according to Colin Purcell, were the single biggest cause of disabling accidents.

The stability of a tunnel roof immediately after blasting was always a big unknown. In one incident in the Murray-1 pressure tunnel, Colin Purcell and a shift supervisor were working together, trying to prise a big slab of loose rock from the roof. The mucking-out team had seen the slab hanging from a clay seam and had been reluctant to touch it, so the two men were trying to dislodge it with crowbars; but it would not budge. As Purcell explained: 'The next shift came on and we told them about it. They must have tried [to dislodge it] and failed, and decided to continue drilling. Two fellows started working on the face and down came the slab, crushing them. It didn't kill them but they would have been better off if it had.'

Joe Bowen, a foreman at Island Bend, was supervising the installation of rock-bolts in the Snowy–Geehi tunnel when the driller felt a big ceiling slab begin to vibrate. The machinery was switched off and Bowen began 'sounding out' the slab with a sledgehammer. It sounded loose so he ordered everyone off the jumbo. The shift supervisor joined him and also had a hit. Nothing happened and he handed the hammer back to Bowen:

> I started on it again and it was definitely looser. A few blows later I felt it shift. I leapt from the platform just as it started to drop. It was twenty-four feet [seven and

Sir William Hudson firing the shot to break through the last rock separating the two faces of the Murrumbidgee-Eucumbene tunnel, 24 May 1960.

a half metres] to the ground and the fall stunned me. As my senses returned I saw the other fellow lying across the locomotive tracks. For some reason I was worried about a train coming, so I crawled across to drag him off. As I reached him the roof over where I had been lying came down, three huge boulders onto the spot I had just crawled from.

While there was danger at every step, Bowen also remembers lighter moments. The language barriers, while they sometimes created safety problems, were, particularly for the Australians, a source of humour:

We were hammering in railway spikes, which were called 'dogs', laying down more tracks for the mucker. Anyway, I could see this Italian chappie looking around, confused.
 'What's up Spiro?' I asked.
 The man shrugged helplessly: 'Shift boss say go outside and get box of dogs—but all I see is fucking cats.'

Spiro's confusion had clearly arisen because of the hordes of feral cats that regularly congregated around the mess at the campsite.

Bowen also remembers the love–hate relationship between the Australians and the Americans in the tunnels:

To the Yanks we all were second-class citizens. They were anti-union and went out of their way to break union rules. It wasn't until Thiess Brothers was on the Scheme that there was any real industrial harmony.
 We worked under one particular Yank, Whitey Dunbar, a supervisor with the Kaiser group. He was a tough character but a real bastard.
 One day I was making up primers outside the tunnel and heard screaming. There had been a rockfall. A Portuguese chap staggered from the dust billowing out of the tunnel opening, holding his arm and screaming. He was wearing wet-weather gear for hosing down dust after blasting. The first-aid fellow rushed up, cut the rubber sleeve and sweater underneath, put the arm in a splint and bound it up.
 Then out of the dust walked Whitey, holding his wrist. 'Better look at my hand,' he said casually. It was smashed to a bloody pulp. He must have been in agony but he never even grimaced. We hated him but had to admit he was a man.

During the twenty-four years of its construction, the Snowy Scheme produced many stories of heroism and horror; but for those who were present none were as horrific as the incidents that resulted in multiple fatalities.

On 16 April 1958, at Kenny's Knob, the winch which lowered the man-haul cage to the Tumut-1 underground power station seized. The cable snapped, and four men went screaming to their deaths, plunging 370 metres into a black hole. The lights at the bottom of the shaft were blown by the impact and a young Latvian doctor, Dr Jon Baksa, had to be lowered alone into the darkness to locate the bodies and pronounce the men dead.

A similar accident had occurred the previous year while the two pressure shafts to Tumut-1 power station were being lined with huge thirty-tonne steel pipes. A massive winch lowered the pipes down each 366-metre vertical shaft. As each pipe was lowered, a steel platform was placed on top of it and several men rode the platform on the slow, careful descent. Their job was to keep the load stable by fending it away from the walls of the shaft.

On the afternoon of the accident, Max Paterick, the Kiandra postmaster's son and packhorse handler who was back with the Authority's transport division after his brief mining stint, arrived with a load of stores. He stopped to watch the activity while he waited for the supervising foreman to sign for the deliveries.

Max watched, fascinated, as a crane positioned the giant pipe into the opening of the shaft where it was hooked to the winch and fitted with the platform. When all was ready, the foreman, an Irishman, told the riggers—a Frenchman and three Italians—to go on down without him. He had paperwork to do. The foreman and Max watched as the winch began to lower the massive load into the hole, then went inside the lifthouse to mark the delivery dockets. Max had just closed the door behind him when there was a loud bang. The pair rushed back outside to see the steel hawser spilling freely off the winch drum.

A distant boom sounded from within the shaft, and the two men sprinted to the edge. The hawser came to the end of its length, snapped, and disappeared into the void. The screams of a man in agony reached them and, peering cautiously over the edge, they could see one of the riggers about ten metres down. He was impaled and slowly spinning on the end of a length of steel reinforcing rod which protruded from the top of a section of concrete lining. The man, recognised as the Frenchman, had been skewered through the middle and was spinning in the rush of air that blasted back from the plummeting load. He was face-down and as he turned his blood fanned outwards, making neat crimson spirals before it fell into the black hole.

Of the huge pipe, the platform and the three other men, there was nothing. The shaft was more than 300 metres deep and the two men could see only a black emptiness, from which there came nothing but deathly silence—a silence which only seemed to amplify the agony of the man just below.

They tried to shut their minds to what would be found at the bottom of the shaft and worried instead about the screaming, tortured rigger. The man had started to clutch at the rod that protruded from his stomach, his arms and legs moving in jerky spasms like the legs of a pinned insect. Max ran back to the hoist house to raise the alarm, call for the ambulance and doctor and look for a bottle of rum or scotch. He desperately needed a strong drink. The foreman began organising workers at the site to weld together a platform with which to rescue the injured man.

It was about half an hour before the platform was ready and Dr Berents arrived from Cabramurra. All this time the skewered man's screams etched deeper and deeper into the minds of the men working frantically to rescue him. When the platform was ready, the doctor and several workers were lowered by crane to the stricken worker. Dr Berents injected him with a pain-killer, then the workers, their own faces deathly pale, pulled him gingerly off the rod. It was a messy, bloody task. The doctor tightly bound the man's torso to stem the blood flow, and they were lifted back to the surface.

Meanwhile, Max Paterick had decided to leave the shocking scene. There was nothing he could do and he returned to the transport pool at Cabramurra. However, his involvement in the drama was not finished. The injured man was taken up to Cabramurra on the back of a jeep for transfer to an ambulance and Max was the only available driver. 'Fast as you can,' ordered Dr Berents.

It was normally a two-hour drive from Cabramurra to Cooma on the narrow, winding road. Max did it this time in fifty-three minutes. About ten kilometres from Cooma the ambulance was met by a police escort. As it sped through Cooma's streets the convoy attracted curious stares, but the townspeople had by now become accustomed to such emergency dashes from the mountains.

After hospital staff had wheeled the injured man away for surgery, Dr Berents and Max stood together leaning against the ambulance. 'I could do with a drink,' said Max. Wordlessly, the pair walked to the nearest hotel where they ordered two double scotches. They disappeared in seconds.

Back at the shaft, salvage teams were organised to descend to the bottom to clean up the mess. It was a sickening sight. Little remained of the men. Two had been dashed to pieces on the drop after being blown off the platform by the speed of the fall; the third had ridden the platform all the way to the bottom but was then shredded by the steel hawser that followed behind.

Accident investigators later reported the cause as a broken winch and failure of the winch's emergency braking mechanism.

The skewered man died a week later in hospital.

In March 1963 Queen Elizabeth II made a three-day visit to the Scheme and was taken into the tunnels. To the workers, particularly the non-British ones, the carefully orchestrated exercise was like a bizarre comic opera. Clad in shiny new yellow oilskins, the workers lined the tunnel while the Queen inspected them and their work. She wore a spotless white oilskin and hard hat. The mud outside the tunnel had been covered with tons of carefully spread and rolled gravel. The normally grimy workers barely recognised either themselves or their workplace.

The Authority had been asked to assemble a guard of honour comprising tunnellers who were ex-servicemen. When the Authority called for former soldiers to present themselves, two Germans turned up: 'Well, you didn't say which army,' they protested indignantly when told to clear off.

The year 1963 is also remembered for another multiple fatality, regarded by some as the most horrible that rescue workers had to confront. It happened at Island Bend on 21 December, the Saturday before Christmas, during a concrete pour in what was known as number-four shaft.

Concrete was being poured 160 metres down a vertical steel pipe to a large bucket, from which it was spread behind formwork to make a lining for the tunnel. At 11.30 a.m. the gate in the bottom of the bucket jammed. As the men used hammers to try to bang it free, workers on the surface, unaware of the problem, kept pouring concrete into the pipe. The concrete quickly filled the bucket and then backed up into the pipe.

Before long there were hundreds of tonnes of wet concrete suspended in the shaft. Without warning, the pipe broke from its supporting brackets and collapsed to the bottom of the tunnel. An avalanche of wet concrete swept all before it along the tunnel—men, machinery and wreckage.

Two men were buried, four were entombed in a chamber formed by wreckage and cement, and another man was stranded in waist-deep concrete. It took ten hours to free the entombed men who, though seriously injured, were alive. In that time the concrete trapping the man buried to his waist had almost set. Under the sickly yellow wash of emergency lighting, doctors and first-aid men had to amputate his legs to free him. He died soon after as a result of his shocking ordeal. The bodies of the two men buried were recovered some hours later. Those killed were two Yugoslavs, Smail Mezlyak, 31, and Jere Coranov, 25, and a Spaniard, Aurelio Mayorol, 26.

Electrical storms occurred almost daily in the mountains, and premature explosions, many of them set off by lightning, were another cause of death. Lightning would strike the ground, find a water seam and travel sometimes hundreds of metres into a mountain to where a tunnel face was being blasted. On striking the underground chamber it detonated any primed explosives.

At the sign of an approaching thunderstorm, first-aid officer Ulick O'Boyle had to monitor a lightning detector. But he 'sweated blood' waiting for the American engineers, secure in their base-camp huts, to heed his warnings. They would delay until the last moment before issuing the order to stop blasting. Often they chose to ignore the warnings, using stalling tactics until the storm had passed. It was a calculated risk, and if there was a lightning strike which triggered an explosion they could always blame the watchman for not impressing upon them the seriousness of the storm.

The men who manned the first-aid posts spent their hours dreading the sudden jarring ring of the field telephone that linked them to the dam sites and tunnels. When there was an accident, they could sense the fright and tension behind the urgent summons: 'Stand by. We need stretchers, doctors, morphine. Dear God, make it quick!'

The scene that greeted medical teams often resembled a battlefield. Ulick O'Boyle, a prolific songwriter who penned many of his experiences for a popular folk group, the Settlers, recalled one incident which moved him to write and record a song entitled 'Big Pedro'.

There had been a rockfall at the face in the Snowy–Geehi tunnel. A big popular Spaniard, Pedro Lafuente, had been hurt; and that was about as much as Ulick knew when the clang clang of the locomotive footbell sounded outside his office. He tucked a stretcher under one arm, grabbed the first-aid box and climbed onto the locomotive's rear platform behind the driver, Gino.

Gino helped Ulick secure the furled stretcher onto the flat engine housing and waited for his nod to re-enter the tunnel. It was eight kilometres to the face. They headed in beneath the roaring blast of the extractor fan under roof lights which disappeared into the utter black 'like a luminous thread of gold'. Gino pushed the hand accelerator all the way forward and the pair bobbed and swayed at full speed to the rockface where the accident had happened forty minutes before.

'Was it just Pedro?' Ulick asked Gino.

'Just him.'
'Is he dead?'
'He no move much.'
'He move at all?'
'I no see. Too many people.' Gino shrugged violently with frustration.

They spoke no more. Ten minutes later they passed the stationary drilling jumbo parked on the track-spur. About 250 metres further on they stopped near the face where the ground was littered with shattered rock. Hushed miners stood on either side of the tunnel staring at the prone figure in yellow oilskins, abandoned like a broken doll.

Big Willie, the supervisor, approached Ulick: 'I see he is dead so I let no one touch him.'
'Thanks Willie.'

Ulick knelt beside Pedro and Gino opened out the stretcher and blankets by his side. They half turned Pedro so Ulick could check and close the empty eyes and lay his ear to the miner's heart. When he lifted his wrist to see if there was any pulse he could feel the bones move away from each other.

Ulick laid the arm down, straightened out the body and with Gino's help lifted it onto the stretcher. They covered it with the blankets. Willie ordered two miners to help lay the laden stretcher onto the locomotive engine again. Gino and Ulick climbed on board and began backing slowly out of the tunnel, Ulick leaving one hand resting on Pedro's leg. As they passed the tunnellers, some crossed themselves, others wept and all removed their helmets and bowed their heads.

By the time they were outside again and had put Pedro inside the first-aid post the area was swarming with engineers, safety officers and inspectors from the Authority, all looking grim and telling Gino to hook up a man-haul to take them inside. Gino didn't need telling. He was already connecting several man-hauls to go in and bring out the shift, which was the practice after a fatality. It was a mark of respect and a necessary break to allow men to recover their nerves before starting again.

The Irish camp manager entered the first-aid centre just as Ulick finished phoning Dr Baksa at Twins Camp, telling him to expect him in the next hour or so. The Irishman called two men in from outside to lift the stretcher into the back of a Toyota four-wheel-drive ambulance. Ulick locked the doors, climbed into the driver's seat and drove off up the twisting mountain track to the medical centre at Twins Camp.

It was now dark and snow whipped across the headlights 'like ghosts fleeing a fitful life, like the spirit of poor Pedro lying there behind me'. Ulick gripped the steering wheel harder and set the wipers to their maximum speed to clear the snow.

Just a few kilometres out from Twins Camp there was a 'thump' from inside the back of the vehicle. The hair stood up on the back of Ulick's neck and cold sweat dropped down his back. 'Thump thump'. The sound came again. Ulick stopped by the edge of the road, pulled the handbrake on hard, walked to the back, opened the doors and climbed in to take a closer look at his passenger. Pedro looked just as dead as before but something was different about him. His forearm had slipped from his chest and his fist sat on the floor. It had been thumping its protest at every pothole. Ulick wiped the sweat from his forehead, tied Pedro's wrists together across his chest with a bandage, then continued his journey. He delivered Pedro to Dr Baksa who, with Ulick as an assistant, did the things doctors do to trauma victims—plugged orifices and cleaned and straightened the body.

The next day nobody went on shift and a lot of the tunnellers drove over to Khancoban to have a drink and forget. On the way home later that night a car full of men went over a cliff, killing three of the occupants. 'They were just dead unlucky,' said Ulick. 'That's the way it was up on the Snowy.'

Luck did play a significant role in the survival of men working at the face, but the odds were regarded as significantly shortened when a man was ordered behind formwork to tamp down concrete being poured during the lining of a tunnel.

It was considered the worst job of all. Australians refused point blank to do it, more than once opting to quit the job rather than follow an order to enter the cavity being filled. Men were sent through a small opening in the formwork to settle the concrete with compressed air vibrators. They worked in darkness in a tight space under a deluge of concrete. A man had only to lose his balance or be hit by a rock in the cascading concrete, and he could be knocked to his knees and buried in moments. No one on the outside would know of his plight until it was too late.

Despite strong denials by the Authority, the story persists today of two Italians who were sent in and never seen again. They were reported as missing at the end of the shift and listed by the contractor as having walked off the job. It would be the worst kind of nightmare and clearly no one would be too keen to acknowledge what really might have happened.

Sometimes accidents made news; sometimes they passed unnoticed by the outside world. On 31 July 1959, the *Cooma-Monaro Express* carried a report about two men who were killed and six who were badly injured the previous day by an explosion three kilometres underground in the Eucumbene–Murrumbidgee tunnel.

A gang had just completed tamping the face and wiring the charges, when one of the thirty-five charges exploded. The blast created a massive shotgun effect, impregnating the men with rock. Those killed were Leonid Harodes, 37, from Estonia; and Frederick Waldon, 28, an Australian. There was no follow-up report about the condition of the injured men, or any mention of the fact that a third man, a young Italian named Ivo Darsie, also died from his injuries. What was reported, though, was the disturbing revelation that the ambulance had to be despatched from Cooma, almost eighty kilometres away, because the contractor had withdrawn the service at Tantangara the previous month as a cost-saving measure.

Explosions which went wrong often caused horrendous injuries in the enclosed space of a tunnel. In the Eucumbene–Tumut tunnel two men were walking towards the exit when the face behind them was blasted. They had not taken cover when the siren sounded because they were more than 200 metres away and considered themselves well out of the danger zone. A slab of rock hurtled along the tunnel from the face and took off both their heads. The velocity of the rock was so great that after decapitating the men it continued on for almost another hundred metres.

Mortally injured men clung to life with desperation, pleading with every nerve against the unfairness of their own impending death. For men looking on, for men still alive, the horror could never adequately be put into words. To them, the dying throes of another was a preview of their own likely end. The only balm against the nightmare was the wet canteen.

On Monday 26 October 1959, another small article appeared in the *Cooma–Monaro Express*, chronicling another fatality, another statistic. The report read:

> A man was killed in an explosion in the access tunnel at Section Creek on Saturday afternoon. He was Ernest Vecchiato, 29, a miner of River Camp.
>
> Police say that about 4.50 p.m. on Saturday, Vecchiato was working a jackhammer in the tunnel. A small quantity of explosive not cleaned from the face exploded.
>
> Vecchiato was severely injured about the lower half of his body and died about 6.20 p.m. in the Sue City medical centre.
>
> Constable Ron Davey of Cabramurra is in charge of inquiries.

The report was straight, matter-of-fact newspaper reporting, the information picked up on a routine phone-around of police stations by the reporter. But like all such accident reports, it belied the personal tragedy.

Ernie had actually been using a hammer drill which he inserted into the remnants of a previous drill-hole. Because such holes could contain undetonated explosives, Ernie's action was clearly against standing orders; but he was also conscious of the pressure for haste. From a space that would eventually be up near the roof of the Tumut-2 underground power station, he was drilling down into the rock which had to be blasted to create the great cavern that would house the installation. In a constant rotation the tunnellers blasted, removed the rubble and blasted again, gradually lowering the floor away from the ceiling space.

Moments after Ernie started to drill, the bit detonated an unfired shot that was still in the rock. He was penetrated from head to toe with slivers of granite. The blast did not kill him instantly and he had to be manhandled out of the tunnel on a stretcher.

'Poor Ernie—he goin' to die,' he repeated in broken English as he was carried from the rubble to the surface where he had to be ferried by jeep to the first-aid station eight kilometres away at Sue City.

His workmates tried to reassure him: 'Cut it out Ernie, you'll be right.' Some of them wept bitterly as they carried him out. Ernie was a popular fellow, expressive and gregarious and liked by all for his humour and his fine voice. After he had finished on the Snowy he wanted to be an opera singer. But his dreams, and the dreams which had brought him to this strange frontier on the other side of the world, were fast slipping from his grasp. He peered pleadingly from his torn face: 'Ernie die; oh, Ernie die,' he said.

His friends went with him to the medical post, encouraging him and willing him to pull through. But Ernie died at the first-aid station an hour and a half later. He had been in terrible pain but had never once complained.

So many had been killed and forgotten that this time the men wanted to remember their mate. With the rock pulled from his body by the medical officers they bordered a small garden outside the medical post and christened it 'Ernie's Rockery'.

Sadly, both Sue City and Ernie's Rockery later disappeared under the top arm of Talbingo Reservoir.

CHAPTER NINE

Eno Grazzi an Italian mechanic and Dieter Amelung above Guthega, 1956.

NEW AUSTRALIANS

By the mid-1950s, the Snowy Scheme, as well as more general demand for labour for industry development, had pushed immigration numbers up to an average of 125 000 people a year. With more than half of these being from non-British backgrounds, it was the beginning of the end for Anglo-Saxon-Celtic Australia. A lot of Australians felt uncomfortable about the increasing number of non-English-speaking newcomers from Europe.

Most native-born Australians remained fiercely British—so much so that until 1948 there was no such thing as an 'Australian citizen'. Before then Australians were still all 'British subjects', even though nationhood had been granted in 1901. A bill to define and give status to 'Australian citizen' was prepared for Parliament in September 1945, but was deferred because of an anticipated community backlash. It was not re-introduced and passed until March 1948, from which time the occupants of this country officially became independent citizens for the first time.

But to most, an Act of Parliament was just a piece of paper and the majority still steadfastly regarded themselves as 'British'. Only twenty years earlier, the people of an entire state, Western Australia, had voted in a referendum to secede from the Commonwealth of Australia and be reinstated as a British colony. Their ambition was thwarted because the pro-secessionists just failed to gain the required two-thirds majority.

In urban Australia, and particularly among the professional class, there appeared to be no real sense of belonging in the country. Many behaved as though their birth here had been one of nature's unfortunate mistakes. At the earliest opportunity they fled back to the embrace of Mother England. The ultimate achievement for some was to be able to go to England, 'make it' and put down roots there in much less hostile soil.

The Australian people's spiritual bond was clearly and firmly with England and all things English. The dilution of British stock in the population was, therefore, cause for disquiet. Yet the 'aliens' were bringing with them the first real seeds of nationhood, the beginnings of a truly independent Australia, which only now, towards the end of the twentieth century, is showing signs of coming to fruition.

In 1955 the government's annual immigration target was 70 000 assisted migrants and 55 000 full-fare migrants—a total of one-hundred and twenty-five thousand. A Gallop poll, held in June and July 1955, asked people if an average of 125 000 migrants a year was too many, too few, or about right. The results were:

	Per cent
Too many	45
About right	39
Too few	10
No idea	6

About this time migrants were being blamed for the first pangs of inflation being felt in the economy, as well as for a statistical increase in crime in Sydney and Melbourne.

However, the actual evidence on both claims was to the contrary. Reports by business leaders demonstrated that without the migrant intake Australia's economy would have been severely retarded. A report published in September 1955 by the director of the

```
eference: 353                              Australian Mission,
morandum No. 268/51                              Bonn.

                                           5th November, 1951.
```

MEMORANDUM FOR:

    The Secretary,
        Department of External Affairs,
           CANBERRA.

GERMAN MIGRATION

The following is a translation of a press cutting taken from the German press of 20th October. Short reports of the statement were carried by a number of papers.

> "Mass German migration to Australia can become a considerable danger for Australian democracy", said the President of the Council of Australian Jews, Green, in London on Friday. "In the light of German rearmament and the resurrection of neofascist activities we in Australia consider it desirable to keep this danger in mind".
>
> The Australian Government had let it be known that Germans would be permitted to enter the country in limited numbers, some six-thousand yearly, and that these would be carefully screened. However, it had been established that immigration is being carried on by the back door. Various enterprises had begun to recruit German workers with a two-year contract. These Germans were not coming to Australia as immigrants, but with labour contracts. In reality, however, they had every intention of remaining in the country".

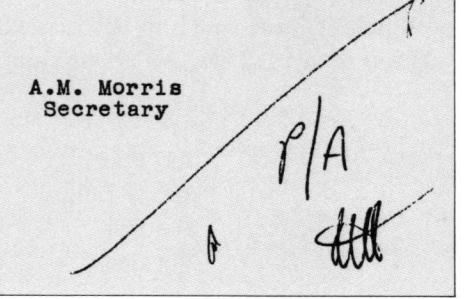

                                            A.M. Morris
                                            Secretary

Outside the Snowy communities, German migrants in particular were unwelcome, as seen from these samples of the vigorous lobbying by various groups.

LETTERS: AUSTRALIAN ARCHIVES OFFICE

26.2.51

7 Ridgeway Avenue
Kew E.4.
Melbourne.

Mr. H E Holt M.H.R.
Minister for Immigration
Parliament House
Canberra.

Dear Sir,

I am sure I am one of many citizens who feel impelled, after hearing the case against mass migration of Germans to Australia presented at a meeting at the Sydney Town Hall last night, to make a protest against this proposal.

I am particularly impressed with the danger of such a large-scale migration of Germans from an age-group which bore the full brunt of Nazi education and propaganda. It seems obvious that this generation of youth was so completely indoctrinated with a philosophy wholly alien to Australians could not in a lifetime adapt itself to our traditions or accept our democratic methods.

The impossibility of adequate screening has been clearly shown by innumerable authorities, and our experience in this country, when it has not been possible even with a comparatively small stream of Displaced Persons to completely exclude Nazis, gives cause for great doubt as to mass migration of any people who have been subject to Nazi training.

As a Victorian Council member of the Australian Legion of Ex-servicemen & women, I can assure you that, notwithstanding public statements by several R.S.L. authorities, the feeling against the proposal among rank-and-file ex-service men is outspoken and profound.

I hope the Government will immediately reconsider the entire proposal.

I am
Yours faithfully
Helen G Palmer

Australian Industries Development Association, Mr C.P. Puzey, said that without the migrant intake Australia's spectacular industrial development between 1947 and 1955 would simply not have happened.

'They have taken on heavy and unpleasant work and work in distant, remote locations which Australians shunned in a full employment economy. They have aided production and given much needed mobility to the workforce,' he said. Mr Puzey also pointedly noted that the skills level of the average migrant was significantly higher than that of the average Australian.

But Australians, whose only link with a distant world was the umbilical cord with Mother England, clung with religious zeal to their xenophobia. The authorities found themselves repeatedly having to delve into statistics to dispel the ever-popular notion that non-British Europeans were responsible for increased crime.

They also had to counter strident outbursts from various sectors about the national mix. The Combined Unions Committee of the railway workshops at Newport, Victoria, for example, repeatedly protested about the German intake. It sought to remind the government that Germany was a former enemy and its people had been 'thoroughly indoctrinated' with Fascist ideas. The committee repeatedly urged the Government to draw much more heavily from British people.

But both government and business continued vigorously to pursue the broad-based immigration program, and through press, radio and newsreels did their best to explain, cajole and shock the community into accepting the need for all kinds of migrants. The government used the spread of Communism in Asia to launch the catchcry 'populate or perish', and warned that Australia's population needed to increase from its existing eight and a half million to at least twenty million by 1979 if Australia hoped to be able to defend itself. It argued that with adequate power and water, Australia's population could reach 120 million by the middle of the twenty-first century.

The government launched a publicity campaign to highlight the impending arrival of the one-millionth migrant. It wanted to make much of the symbolism: 'Will he or she be a Brit, Pole, Hollander, or Italian?' the promotion asked. But it was a rigged event.

Secret minutes of the Commonwealth Immigration Advisory Council meeting of 7 and 8 July 1955 reveal that, after some discussion, it was decided not to let the choice of the millionth migrant be left to chance. It was considered imperative the person be British. For maximum coverage it was also deemed preferable that the person be destined for Melbourne, rather than Fremantle.

The Department of Immigration's new publicity section went one better and decided the best 'selling' opportunities lay in the millionth migrant's being an attractive young woman—someone who 'typifies the young, eager and enterprising migrants who have set forth to apply their energies and skills to building a new life in this land of opportunity'. The chief migration officer in London was put in charge of the search and chose a 21-year-old stenographer from Yorkshire.

The British priority was reflected in more than the government's policy salesmanship. For example, British migrants were given far superior hostel accommodation. While British families enjoyed comparative luxuries such as sitting rooms, the space available for non-British families was so restricted that families had to be split. Children were often forced to stay behind in the initial holding centres until the family breadwinner was able to secure permanent outside accommodation for the whole family.

When the Department of Labour and National Service sought to have the dependents of British nominees treated in the same way as non-British migrants, its advice was promptly rejected by the Advisory Council. Acceptance of such a proposal, the council claimed, would adversely affect recruitment in the United Kingdom and arouse public criticism: 'We could not have in operation a scheme where some British families were divided, while others were accommodated in family units in hostels,' it said. Yet it had no qualms about non-British families being put into that situation.

By 1954 and 1955, however, this policy was beginning to backfire. A report to the Advisory Council in April 1955 warned that British migrants were furnishing their hostel accommodation 'in such a way as to indicate they had no intention of vacating'. The report went on to point out the difficulty of explaining to non-British migrants that they could not bring their dependents to live with them when there were vacant bedrooms reserved for incoming British migrants and British migrants in the same hostel who had empty space which could almost accommodate another family.

Although the council still postponed taking action the accommodation situation soon became so critical that the government acted arbitrarily and discontinued the provision of sitting rooms. It also put a two-year time limit on hostel accommodation. This drew an immediate and sharp criticism from the New South Wales Trades and Labour Council and an organisation called the National Council of Women.

Another issue to emerge was the state of mind of some of the European migrants, particularly those who had suffered severely in the war and been the victims of atrocities. As early as 1950 the Department of Immigration reported it was becoming increasingly worried by the incidence of severe mental and emotional disorders among migrants who had come to Australia as displaced persons. It sought a directive from the Director-General of Health and raised the possibility of deporting such people back to their country of origin—a rather perplexing proposal when one considers that many of them were stateless.

The Director-General of Health investigated the matter and, while conceding that some people had lied about severe emotional stress or history of mental illness in order to qualify for migration, he recommended that such cases should be regarded as unavoidable accidents, rather than excuses to send people back. He pointed out further that: 'Migration itself, with its uncertainties, strain and nervous tension, will precipitate some mental upsets.'

There was also concern in the 1950s that migrants coming to Australia from the Eastern Bloc as part of the family reunion program might be bringing with them 'Communist indoctrination'. The first such case involved a Greek youth who had been separated from his family during the German invasion of Greece in 1941. It was discussed at length by the Advisory Council.

The youth was among several children from a village who were taken first across the border into Yugoslavia for safety, and later to Czechoslovakia. In 1954, when Iron Curtain countries began to allow people to leave to rejoin families, the youth applied to come to Australia where the rest of his family was settled. His application was rejected by the Australian Security and Intelligence Organisation (ASIO) on the grounds that he would by then have been 'incurably' indoctrinated with Communism. Worse still, said the ASIO report, it had information that the youth's mother was descended from a Bulgarian family and during the war had declared herself 'for Bulgaria' and had collaborated with the Bulgarian arm of the KKE (the Greek Communist Party).

The Immigration Advisory Council pondered the matter at great length. Then it took the surprising step—in view of the general paranoia over Communism—of rejecting the ASIO report. The minutes of the council's own secret discussions reveal that its members doubted the objectivity of ASIO's approach to the matter and issued a recommendation to the government that young people in such circumstances should be admitted to Australia 'despite the possibility of them having been indoctrinated with Communism'.

Many migrants were themselves worried about Communism, and in particular about the Cold War in Europe. They were anxious about family and friends who were left behind, and put pressure on both the Snowy Authority and the Australian government to allow their families to emigrate.

The following article appeared in the *Canberra Times* on 17 May 1951, under the heading 'Migrants at Snowy Fear for Families in Red Europe'. It showed that for many, even in remote work camps in the Australian Alps, the troubles at home were ever present:

> PERISHER CREEK. Wed: Forty four Germans and other European migrants, at a meeting last night, decided to petition the Minister for Immigration, Mr Holt, seeking his aid to bring their families to Australia.
>
> Working on the Snowy Scheme, the migrants asked that their names should not be used because they said: 'We live in daily fear of reprisals against our families if we make statements here in Australia.'
>
> One man with a brilliant war record in the Russian Army when it fought on the Allied side, said that because he had made a statement to broadcasting and press representatives there was little left in life for him.
>
> 'Last week I received information that my sister-in-law committed suicide because of statements transmitted from Australia to the Russian Zone.
>
> 'My own wife says that unless I am able to get her speedily from the Russian Curtain, she too will commit suicide.'
>
> A German said that if the Minister was unable to help, an approach would be made to the United Nations.
>
> He said they had supplied the Department of Immigration with a list of menaced families in European countries.

In an effort to allay any fears over Communist influence in Australia, the Immigration Advisory Council conceived the idea of a charter, or statement, of Australian citizenship, which would proclaim allegiance to Australia and signify an understanding of the privileges and responsibilities of citizenship. The council wrestled long and hard with a draft proclamation, which it believed naturalised citizens should sign, but finally threw the matter into the 'too hard' basket. It decided any such emotional or sentimental statement would be treated by the average native-born Australian with suspicion or ridicule and rendered meaningless.

But, despite the scepticism with which some Australians regarded immigration, the program was being watched from overseas with interest and respect. It had become the largest sustained effort of its kind in the world. By the mid-1950s, when the program was still in its infancy, Australia had absorbed proportionately more people of foreign birth than did the United States in its peak period of immigration at the turn of the century.

The contribution that New Australians were being encouraged to make was considered remarkable by many foreign observers—especially considering that the world had just passed through a period of history marked by extreme racial intolerance. The speed and ease with which most migrants were assimilating into the community was regarded by some as little short of miraculous.

For tens of thousands of these New Australians, the Snowy was their first meal ticket. By the late 1950s the Snowy Scheme was employing up to 10 000 workers, most of them migrants, at any one time. At the end of their two-year contracts with the government, all were free to return to their home countries. Few did.

During its twenty-four-year construction period, the Snowy became a nation unto itself, with Cooma—or perhaps more accurately, Cooma North—its capital. Lifestyles forged by isolation, hardships and the diverse cultural mix gave the Snowy people a strong sense of identity. The experience made an indelible mark, which all still carry today. For many, their years on the Snowy remain the highlight of their lives.

Dieter Amelung, the East German refugee electrician, was a typical example. After the bitter disappointment of his first night at Kenny's Knob life improved daily. On his second day in the mountains the foreman from Cabramurra returned to collect him and put him to work in the small diesel power station that had been built as a temporary

Dieter Amelung, Cabramurra, 1959.

The Cabramurra wet canteen, 1954. Dieter Amelung is second from the right.

measure for the town. Working again in his trade was a joy in itself, but a week later, after he collected his first pay packet, the world was positively rosy. He received four times the amount he had been earning in Germany.

As time passed, Dieter, like many of the migrant workers, found the loneliness an oppressive burden but found comfort in the bush. Its wilderness was both a refuge and an adventure. He began to spend more and more of his free hours bushwalking. He bought a rifle and an old motor bike and rode regularly to the Yarrangobilly Caves where he hunted in the steep valleys for rabbits. This in itself was a thrill. It would never have been allowed in Germany.

About a year after his confusing arrival, Dieter learned for the first time of the role the Authority had in mind for him when he was recruited. The Guthega project was nearing completion and Dieter was informed that he would be offered a position as an operator on this, the first power station to begin generating electricity from the waters of the Snowy.

When Guthega was opened on 23 April 1955, signalling to the country that the Snowy was up and running and nothing would now stop it, hundreds of workers from all over the Scheme flocked to the ceremony to cheer boisterously every word uttered by the Prime Minister, Mr Menzies.

It was a big occasion. Power station operators had spent weeks checking and test-running the system. In the days before the opening they were also put to work making sure the station and its grounds were spotless. On the day of the opening everybody was in a festive mood and even Menzies, whose support until now had never been certain, picked up on the atmosphere.

The ceremony was performed on a dais, erected outside the station above the outlet pipes which lead to the Snowy River. The Prime Minister ceremoniously turned a handle which released water from penstocks to set the massive turbo-generators spinning. He then entered the power station, where he was directed to a switch which put the turbines on load. He faced the visiting dignitaries and the cheering workers:

> This is a very dramatic and exciting occasion. We are here today looking at the first fruits of the Scheme which represents one of the greatest efforts of co-operation in the history of Australia.
>
> The water that will be conserved and made available by this Scheme by the time it is completed will alter the face of a vast area of Australia. The Scheme will contribute more to Australia than any other single enterprise in the entire history of the continent.
>
> I am now going to perform a mechanical act; and when I perform a mechanical act, heaven knows what will happen.

The workers laughed and cheered more.

> I'm going to move a switch, but as usual when I move a switch nothing will happen for a while.

As the Prime Minister turned the handle, the sudden roar of water rushing into the turbines drowned out even the tumultuous cheers.

> They appear to be at work, so I declare this power station open.

Standing in the background, to make sure the Prime Minister pulled the right levers, was a young man in a new grey dustcoat. He was Dieter Amelung. To him, the Prime Minister loomed as a big, benevolent father figure. 'Dieter and Mr Menzies opened the

Dieter Amelung, Guthega Power Station, 1955.

power station,' he proudly told friends after a front-page photograph in the *Sun-Herald* showed Dieter in the background behind Menzies.

The opening was not without its drama. For the purpose of the ceremony, it was arranged with the New South Wales Electricity Commission for the station to generate thirty megawatts, reaching that figure slowly in five megawatt increments to give the coal-fired steam stations time to back off. But the press photographers kept asking the Prime Minister to repeat the action of turning the lever. Menzies was happy to oblige, but in doing so he rapidly pushed the lever to the full thirty megawatts. 'The Sydney operators were on the phone screaming, but what could we do? It was the Prime Minister who was causing the trouble,' said Dieter.

The opening of Guthega wasn't the last time Dieter was to come face-to-face with the nation's leader:

> One day I was in Canberra on my way to Sydney when I asked a taxi driver to show me the sights. We drove off and I gazed about as he gave a running commentary: 'Over there's the American Embassy...and over here is the Lodge where the Prime Minister lives...'
>
> I wanted to see where Mr Menzies lived so I asked the driver to stop. I got from the car to take a photograph. As I stood there with my camera another taxi arrived and stopped up at the gate. To my astonishment out climbed Mr Menzies and posed while I took his picture. He then climbed back into the taxi and it disappeared through the gate.
>
> I couldn't believe it and my taxi driver was speechless. The Prime Minister must have thought I was a press photographer.

A few months after the opening of the first power station, the man most responsible for getting the Scheme off the ground, Nelson Lemmon, quit politics.

In the 1949 election, when Labor lost to the Menzies Liberals, Nelson Lemmon also lost his seat of Forrest in Western Australia, which he had held for seven years. He resigned himself to a return to farming, but under Chifley's persuasive influence sold his Ongerup property and moved to New South Wales where Chifley promised him a safe seat.

Although Chifley died eighteen months after Labor lost office, the promise was honoured, and Lemmon was given the seat of St George, in Sydney. He returned to federal Parliament in 1954 amid press speculation that he was a likely contender for Labor leadership. Some pundits even picked him as a future prime minister. By the time Lemmon had taken his new seat in Parliament, however, the Australian Labor Party had entered the wilderness it would occupy until the Whitlam-led return in 1972. It was not a happy time for Lemmon, and he regretted his decision to remain in Parliament. A year later he turned his back on politics forever.

For William Hudson, the man Lemmon selected to make the dream a reality, the opening of Guthega lifted a great weight from his shoulders—though he was to feel a gentler pressure there two months later, when he was knighted.

Sir William Hudson was clearly proud of the role the migrant workers had played and of the way they had approached their work in such gruelling conditions. In an address to the New South Wales Good Neighbour Council in 1958 he said:

> We made mistakes, but have taken care not to repeat them. In the early days we recruited 750 German technicians, whom we segregated into separate camps. They could not speak our language. They were lonely, strangers in a strange land. They became restive and trouble started. When we saw the error of our ways and placed them with Australians and other new Australians, the trouble disappeared. Today most have settled in Australia with their families.

The welding together of so many people of such diverse nationalities did not happen overnight. Some Australians were openly hostile to the non-English speaking newcomers, particularly in the Scheme's early days.

One afternoon in midwinter 1951 Jock Wilson, who had left Three Mile camp and become a transport foreman at Adaminaby, was phoned from Cooma and ordered to take a Snowy truck to Kiandra and offload two supply trucks from Cooma which were not equipped with chains to go all the way to Tumut Pond. Jock took with him another driver who wanted a lift to Cooma in one of the returning supply trucks. They met the supply trucks at Kiandra in the late afternoon. A storm was brewing and, by the time the stores were offloaded onto Jock's truck, dusk was closing in and it had begun to snow heavily. Everyone was in a hurry to start moving again. Jock headed towards Tumut Pond with his cargo of provisions, while the other drivers turned to drive back to Cooma.

Jock had only gone about a kilometre when he lost sight of the road and his truck sank into a deep snow drift. There was no way to free the vehicle so he trudged on foot back to a small survival hut near the Kiandra ski chalet. When he arrived he found it already occupied by a bulldozer driver, who had also been caught in the storm, and the men who had been returning to Cooma. Both trucks had struck early trouble. One had blown a con-rod through the side of its motor and the other had slid off the road.

The survival hut had only a small amount of wood for its small stove and no food. Jock knew the proprietor of the chalet, so he struggled across through the mounting blizzard to seek accommodation for the five men. He did not mention to the proprietor that the others were Europeans. It did not occur to him that it would make any difference. 'She'll be right,' the proprietor assured him.

Jock returned to the hut and told the men the welcome news. They all trudged across to the chalet. As they shook themselves down in the entrance area, the proprietor called Jock to one side: 'I can give you a bed, but not these other fellows.'

'What do you mean?'

The man hesitated: 'Well—look at them.'

Jock did. They seemed no grimier than he was, especially considering one had been in an accident, another had been under a broken engine trying to find out what the trouble was, and the bulldozer driver's clothes were probably permanently oil-stained. 'That's nothing. Once they're stripped off they're okay,' Jock protested. But the proprietor had made up his mind: 'No way. They're not coming in here.'

It dawned on Jock that it was probably more than just their grimy appearance which perturbed the man: 'Well if that's the case, you can stick it. I'm not coming in either.'

The men marched back to the hut where they were stranded for five days. Their only food was cornflakes, condensed milk and dried fruits, which they took from a crate on Jock's truck. The men huddled in the hut around the meagre warmth from the small stove, while a howling blizzard plunged the outside temperature to well below zero. On one occasion four men had to go outside to rescue one of their number who had walked just forty metres away to go to the toilet. The cold had struck him down like a knife. When he was dragged back into the hut he was encased in ice. All the while the men were tormented by the nearby lights of the chalet whose warmth and shelter had been denied them.

Ultimately they were rescued, but even then they were made to suffer. Each day Jock had trekked to the Kiandra post office and had telephoned Cooma to inform staff there that they were still stranded. But no one in the Cooma office saw fit to radio the information to Tumut Pond, where search parties were out every day looking for Jock and his supplies. It was assumed that, blinded by the blizzard, he had driven over the edge of the access track.

On the fifth day of the storm, and with no end in sight, Jock decided to try to ski to Tumut Pond for help. He was barely half-way when he met a snowcat on a search sweep.

When he finally reached Tumut Pond with the supplies Jock was sacked for failing to inform the camp about what had happened. He and the men with him also had their wages docked to pay for the food they had consumed from the truck during their ordeal. Jock's suspension remained in force until the engineer in Cooma, whom Jock had been telephoning from Kiandra, verified that Jock had indeed told him what had happened. Even then Jock had to stand over the man while he made the call.

The massive influx of migrants into a concentrated area, such that they became a dominant part of the population, often proved a difficult test of Australians' confidence in themselves.

This could be seen in the obvious resentment displayed by young local men towards the large numbers of migrant workers who came into town every weekend. The young migrants had healthy pay packets and treated the local girls with a stylish respect. The Australians, not known for their deference to the opposite sex, often found themselves outmanoeuvred and outclassed. Their only allies were local parents who tried where possible to prohibit their daughters from mixing with the foreigners. Girls who did date migrants were often branded by the local community as sullied. Over the years, however, quite a number ignored the threatened stigma and married workers from overseas.

The lack of Australian confidence was also shown in the way the Authority, like the whole country, clung so doggedly at first to its British roots. The Snowy Authority in many ways provided a window into the future for the wider community. The first turbines and generators it installed were all of English manufacture—the decision was automatic. Twenty-three years later in Tumut-3, the last power station to be built, they were all Japanese.

To begin with, the Authority's management was firmly pro-English in its bias. In the main it refused to recognise qualifications from other than Australian, British or

American universities and institutions, an attitude that severely hindered, and even destroyed, the careers of many migrants.

The canteen manager for many years at Island Bend was a Polish criminologist. He was forced to take up this less than exacting occupation because Australia did not acknowledge his qualifications from Warsaw University. When Dr Jon Baksa, the young Latvian doctor who was lowered into the shaft at Kenny's Knob after the tragic man-haul accident, had first presented his medical credentials, he was brusquely informed that he might be allowed to work as a hospital orderly. To become an Australian doctor, he was told, would require ten years' training and would cost 10 000 pounds.

Baksa opted out and became a labourer. Eventually, though, commonsense by the authorities, prompted by the desperate shortage of doctors in the mountains, enabled him to take a retraining course and to practise—much to the gratification of the isolated mountain communities where he took his skills.

Sometimes, however, those who managed to break down the barriers ended up in a kind of social twilight zone.

In an unusual move for the time, the Authority in 1955 appointed a woman, Ksenia Nasielski, an Estonian who spoke five languages, as its senior employment officer. But as a member of 'staff' Ksenia found herself in an anomalous situation. As she was not an engineer, she did not qualify for a house with a verandah; as she was not an engineer's wife, she did not qualify socially for the tea parties.

It did not particularly bother her. Indeed she was often amused by the situation, especially when she compared it with the much more relaxed atmosphere she had known in the mountains during her first years with the Snowy.

Ksenia was born in Leningrad in 1920. Her father was Estonian and her mother Russian. On the train on which the family moved to Estonia, after it had become an independent republic, she was the only child who survived the journey. The rest perished from typhoid.

Estonia was granted its independence as a gesture of appreciation by Russia after it had been one of the first republics to recognise the new Union of Soviet Socialist Republics. The family lived happily there until 1941 when Estonia was reoccupied. Ksenia had been studying at Tartu University, one of the oldest in Europe—it was founded in 1632—when the Russians marched in again.

Because her father was a judge and a potential opponent of the occupation her parents were sent to Siberia. Ksenia had to leave university to find work and look after her father's small farm. She expected her parents would return one day. They never did. Their fate and the way their lives had ended could only be imagined.

In 1944, when Estonia became a battlefield, Ksenia fled to Germany. When the war ended she found work with the British occupation authority in Hamburg. She was stateless, her only home being a refugee camp where she met her husband, Adam, who had fled from Poland.

Neither of us had surviving relations in our home countries so we had to start again and wanted to do so in an English-speaking country. We were considering USA, Canada or England.

But at a party we met a Mr Grey, who was recruiting people for Australian work programs. All we knew about Australia was that it was far away and had a lot of sheep.

He told us: 'We don't need white collar, or even blue collar workers; we are looking for people prepared to do manual labour—women for nurses, housemaids and factory hands and men to work on the farms and in factories.'

So we came to Australia, to make a new country our new home. I got work as a housekeeper for a doctor in Sydney and Adam worked for the Department of Main Roads. At the end of our two-year contract in 1951, we came to the Snowy. Adam was offered employment as a junior catering officer at Island Bend and I worked in the office making up his orders.

The air was fresh, the sky beautiful and we were surrounded by mountain walls.

In summer we went for long walks and sometimes swam in the Snowy River; but even in January it was very cold. The bush with its birds, heath and alpine flowers was new and something I had never seen before.

In winter it was harder. I didn't ski because I was too frightened I would break a leg. It would mean an agonising ambulance journey to Cooma, then a train to Sydney; and when you came back you had the problem of having to move about on crutches in several feet of snow.

The couple had to live in separate barracks, but did not mind because it was a start. All around them was bustling activity—the signs of a nation growing and building. Most of the other workers at Island Bend were still living in tents and in winter it was bitterly cold.

You could have a hot bath in the laundry block, but then you had to run back to the barrack through the snow, climb into bed and try and get warm again.

The place was so alive, bulldozers and trucks everywhere. Prefabricated houses were brought up from Cooma in 1954 and it became a little town. I no longer had to drive for one and a half hours in a jeep to Jindabyne to do the shopping.

To some, the women are the Snowy's unsung heroes; particularly the young wives of migrant workers, plucked from homes and families and familiar cultures and brought across the world to build new homes in an alien environment.

Many came from large cities to be confronted with the isolation of primitive townsites. Most were unable to speak English, and their husbands were away working long hours for six and sometimes seven days a week. Often they were frightened by the bush noises and the blizzards and all the men around them, but they had no one to turn to. They simply struggled on, uncomplaining, in the face of their fears, the isolation and the extreme climatic conditions.

The first organised support for the migrant women came from one of Australia's oldest institutions, the Country Women's Association. Under the driving force of Merle Mould, a young woman from Middlingbank, down on the Monaro, the CWA moved in to try to improve the lives of the construction town women.

The CWA put libraries in the camps and towns, started English classes for the women

Island Bend township, 1953. The town was the Authority's base for the Guthega and Island Bend projects. Houses were only for senior workers and 'staff'.

and also browbeat the contractors and the Authority into providing rooms and equipment for child health centres. When Merle Mould first raised this with the Kaiser consortium she was shown the door. As she left, she promised that the *Women's Weekly* magazine would be up the next week to see how the Americans treated their workers' wives. The following day, the senior American engineer telephoned to offer a car, a driver and a house for the first child health clinic at Happy Jacks.

Because there was nowhere for the women to go and little for them to do, Merle started an international club; but it soon became the exclusive domain of the engineers' wives, many of whom were well educated and starved of intellectual stimulation. One day Merle complained that the club was not reaching the people she intended. 'Oh, but they're only the workers' wives,' she was told. 'Yes, and they're the ones I worry about,' she responded.

Because of the dangerous work and the isolation of the towns, the people who inhabited them lived life to the full. They also relied on each other for such social activities as film nights, stage shows or even darts and cards in the wet canteen. As a result the social distinctions in the mountain camps and towns were much less rigid than they were in Cooma. There the barrier between Cooma North and Cooma East, as the workers' section became known, was inviolate. There remained in the mountains an official barrier between wages and staff, but in time the social demarcation became almost invisible. When people needed each other to create a community their nationality or employment status became less relevant.

Even so, life still looked a lot different when seen through the eyes of an engineer's wife: 'We were young and we danced and partied hard at every opportunity. We would go out into the snow in strapless ball gowns and feel no pain,' recalled Betty Mattner, who thought she had come to the end of the earth when her husband, Dick, brought her and their young baby from South Australia to Tantangara in 1958. Dick had secured work there as a tunnel engineer with Utah on the lower Tumut section of the Scheme.

As the wife of an engineer, Betty had a house to move into and did not have to endure separation in barracks. But when she caught her first glimpse of the fledgling town, the advantage of being an engineer's wife was not immediately apparent. Betty had never been in the mountains before and did not know what to expect. She certainly was not prepared for the stark settlement of unpainted prefabricated houses that greeted them at the end of a slow, hot journey on an unsealed road from Talbingo and along the then rather presumptuously titled Snowy Mountains Highway.

Dick must have seen the look on her face. 'Don't worry, they've all been painted on the inside,' he told her. They rounded a corner and he pointed to one particularly ugly dwelling sitting on high wooden stumps. 'That's ours,' he announced proudly. Betty looked up at it in horror: 'I can't live in that.' Her husband was taken aback: 'You've got to, Bet. That's our house.'

The builders had yet to give the house a porch and steps and the front door was nearly three metres off the ground.

'But how do we get inside?'

'That's okay. I'll find a box,' said Dick. Dick was fired with enthusiasm for his work and for the Scheme and was tolerant of its imperfections. He did not see the structure in front of them as an unfinished building, but rather as their home, with all the ideals which that embodied. He found several large crates, stacked them and climbed to the front door, brandishing the shiny new key he had been given. But the door would not open.

Betty watched him clamber down and disappear around to the back, where the door was only about one metre off the ground. Betty looked around in despair. She felt like following the example of the baby in her arms, which was wailing loudly.

Dick found another box to climb on to open the back door, which opened without any trouble. Inside they found the removalists had stacked all their furniture against the front door. It seemed a strange thing to do, but at least it prevented them from rashly opening the front door, and dropping to the hard ground below.

The rest of the house was completely bare: no floor coverings, no curtains, no light fittings, no cupboards and the builders had left the kitchen sink and bath full of dirty brown water and cigarette butts. Betty dragged out the mattress from their stacked belongings, dropped onto it and cried.

It was three months before the porch was added and another month before the steps were in place. In all that time they had to enter and leave the house by the back door, using a wooden box as a step. With an armload of washing or carrying a baby, it was not easy. It was six months before any cupboards were put in the house.

Soon after that, however, things began to improve. One day workmen arrived unexpectedly to build a fence around the yard. When this was completed and painted, the Authority issued roses, lupins and lawn seed to the occupants of each house. After a year, Tantangara was starting to look like a typical country town; but it retained its distinctive 'Snowy' quality.

At Tantangara the wives formed the fire brigade because the town was eleven kilometres from each of the two construction sites—the dam in one direction, and Providence Portal (the Murrumbidgee–Eucumbene tunnel entrance) in the other. It was too far for workmen to attend a fire. The women were rostered and called to action by a siren on the Utah safety officer's house. As soon as they were assembled, they sped off in a four-wheel-drive fire unit. Despite their regular and enthusiastic training, it took at least four of them to hold the nozzle of the hose. The only time they had to contend with a real fire was when a practice run got out of control and they almost burned down the women's barracks.

One of the more endearing personalities in Snowy folklore was Georgina McQuade, the accommodation officer at Cabramurra. As an official hostess she entertained VIP visitors, including royalty, but it was for her evening attire in the mess that she became famous.

On the day she arrived in the settlement it was snowing heavily and the wet canteen was packed elbow to elbow with hard-talking, hard-drinking workers, boisterously unwinding at the end of a shift. The atmosphere was thick with noise and tobacco smoke. Suddenly into their midst glided an apparition in a long flowing dress and stiletto shoes. At the sight of Georgina, the men were struck dumb. A pin falling would have been distinctly heard.

Georgina believed a lady should maintain her standards, whatever the odds. She dressed up for every occasion, dyed her hair every possible colour, and was nicknamed 'The

Duchess'. She dressed formally for each evening meal, crossing the snow and mud between her hut and the mess in gumboots, with her high-heeled shoes tucked under one arm. Once inside, it was gumboots off and high heels on.

Her quest for elegance in adversity set an example that others followed. Engineers started arriving at social functions in dinner jackets (but still in gumboots) and their wives would appear in the latest city fashions, turning even a celebration for a tunnel breakthrough into an evening of *haute couture*.

Georgina was one of only three unattached, single women employed by the Authority in the regions, as areas outside Cooma were known. She did not remain unattached for long. In the first wedding held at Cabramurra, she married Les Neely, a construction worker.

'The Duchess' remained with the Authority for twenty years, and retained her fondness for dressing up. For one royal visit she wore a pink nylon wig and a shift with shallow neckline under a flowing black cape. The local paper reported she had emerald studs in her ears, green eyelashes, green eye make-up and green fingernails.

Despite the constant strain of having to cope with an unfamiliar language and culture, the migrant workers were high-spirited and easily matched the Australians with their own brand of barbed humour. The Czechs had a great love of a particularly strong Stilton cheese, about which the Australians persistently complained.

A group of Czech hydrologists based at Cabramurra one day hid a piece of the cheese in the drill boss's office while he was away. When the man returned two days later they had almost forgotten about it. But their attention was drawn to a commotion at the office and men being summoned with digging equipment. The smell in the room had apparently been so intolerable that the supervisor, convinced that a body had been buried beneath the building, had ordered the foundations to be excavated.

German workers took great delight in accosting unsuspecting Australians, particularly foremen and supervisors. A German would grasp an Australian around the biceps and exclaim in mock awe, 'Strons, Strons.' Then, turning to his mates: 'He strons, eh?' The Australian inevitably smirked self-consciously, puffed out his chest and continued on his way with a self-satisfied smile—though slightly bemused by the German's hysterical laughter. *Strons*, the Australian would later learn, was a German expression for a big, hard turd.

Neville Phee, one of the first operators at the Guthega power station, well remembers his first night in a barrack at Island Bend in which he was the only Australian among a group of Germans, Czechs and Yugoslavs: 'When I entered the barrack, the leader, a German fellow named Bill Schafer, invited me to drink with them. But there was no beer—just brandy, vodka and scotch, which they were all drinking from middy beer glasses. I was given one filled with brandy.'

Neville almost choked on the first mouthful and managed to pour some of it surreptitiously down a sink. But the moment he put his glass down it was refilled.

'Do you want some food?' the German asked. Neville nodded. The top drawer of a clothes locker was opened to reveal a smorgasbord of black bread, salami, cheese, socks and underwear.

'Do you like chilli?'

Neville felt it was only polite to accept their hospitality, even though he had never actually eaten chilli before. He took a healthy bite of the proffered spice. Almost instantly he was gasping and reeling as the roof of his mouth seemed to ignite. Instinctively he grabbed the glass of brandy and skoaled it: 'I don't remember much after that, except I suffered from the worst hangover in my life the next day.'

The following evening Bill Schafer approached Neville again: 'Come have a drink.'

'Not bloody likely.'

'Come on, come have a beer with us.'

'A beer? Last night you said you didn't have any.'

The German grinned devilishly: 'That was last night. We were just checking you out. Tonight you can drink beer with us.'

Neville's third day at work was equally painful, as he suffered from the second-worst hangover of his life. Australians prided themselves on their drinking prowess, but it was a rare event when they outdrank the Europeans.

The many nationalities and the need for the whole community to contribute towards devising its own entertainment combined to create a stimulating environment. People were always learning something new and seeking something different. To the great delight of the inhabitants of Sue City, the main town for the Tumut-2 project, a ballet dancer was found among the workers.

The dancer, Jon Cambell, had been a member of the then prominent Polish–Australia Dance Company in Sydney and had trained in London under the Russian star, Tamara Karsavina. After sustaining a serious knee injury he put aside his codpiece and pumps, took up a meat axe, and was now working as a butcher in the Sue City supermarket.

Sue City, named after the wife of the contractor, Edgar Kaiser, comprised about a hundred houses, a post office, bank, the first American-style supermarket to be opened in Australia and a community hall. The hall was an impressive concrete structure which included a stage, dressing rooms, kitchen and projection room. It was built by the Americans as a venue in which to celebrate the fourth of July. The hall was a short distance from the entrance to the massive tailwater tunnel being excavated to return water from the Tumut-2 underground power station back into the Tumut River, about eight kilometres downstream.

Two of the American women in the town—Carlin Wahley, the daughter of Edgar Kaiser, and Del Blhem, wife of an American supervisor—had been involved in amateur cabarets at home and had been keen to stage a revue for some time. When they learned the young butcher had been a promising ballet dancer they decided the time had come to bring culture to their little settlement at the bottom of the steep gorge below Section Ridge.

Because the Sue City revue would have to cater for a wide range of cultural tastes, Jon was enlisted to choreograph the unlikely combination of a French cancan and Tchaikovsky's *Swan Lake* ballet. The choice was actually dictated by Carlin Wahley's record collection. Carlin and Del Blhem agreed to organise the rest of the women for a cancan troupe, while Jon and his boss at the supermarket, Don Patton, went in search of ballet dancers among the clodhopping miners and construction workers.

At the suggestion of Don Patton, it was decided to present the revue as an hour-long feature midway through a ball. As at most Snowy balls there would be several hundred men vying for dances with a handful of women; but this never seemed to dull anybody's enthusiasm. Once, at Island Bend, almost 300 Norwegians working on the Guthega project flocked into the camp after hearing that five busloads of nurses were coming from Cooma. The five women at the dance had a wonderful time.

After an exhaustive campaign that involved a combination of strong-arm tactics and blackmail, six 'swans' were eventually recruited for the ballet. The camp bosses gave the 'volunteers' two days off in which to be trained in the art of classical dancing.

Accepting that improvisation would be his only salvation, Jon had the camp's carpenters and painters construct an all-purpose Parisian backdrop. He also decided to begin with the ballet's lake scene, which he illustrated by placing a large Indian canoe, brought out from America by Carlin, in the centre of the stage.

The big moment duly arrived and Jon, resplendent in a tuxedo, which had been hired from Sydney along with the cancan costumes and six of the largest tutus available, faced a hushed and expectant audience. Hundreds of workers and engineers from throughout the region had turned up, lured by word-of-mouth advertising—which had omitted one or two salient details about the production. The eyes of many of the workers were glued to the stage, waiting for their first glimpse of lithe young ballerinas in their skimpy dresses.

At the end of an introduction, in which Jon fired the imagination of the assembly with a glowing testimony to the cast, the hall was darkened. He discreetly put the first record on the radiogram.

As beautiful music swirled through the hall the spotlight hovered over the canoe. Disbelief turned to hysteria as a great hairy leg rose from the vessel, followed by the beaming smile and thick bottle-glass spectacles of Igor Ashnikov, a White Russian surveyor, well known in the region for his buffoonery. Before the audience had time to recover itself, the six swans made their entrance—a blurred vision of tattoos, muscles, armpits, tutus and miners' boots.

The audience collapsed into hysterics. Even the men who had been hoping for a glimpse of female thigh forgot their disappointment. The only people in the building with dry eyes were the 'swans', who stomped through their paces stony-faced with embarrassment.

When the music finished and the ballet dancers had scurried into the wings to change and make their escape into the anonymity of the night, the cancan troupe strutted its stuff; this time giving some of the workers a glimpse of what they had originally turned up for.

Word of the revue spread rapidly through the regions, with requests from all the camps and towns for it to go on tour. But it was a once-only performance; nothing was going to lure those 'swans' back on stage.

Occasions such as the Sue City revue were a welcome tonic, especially for the single migrant men, who often suffered badly from homesickness and loneliness.

Many of the Italians sent a large percentage of their pay packets home to Italy to attract girlfriends and fiancees out to Australia to join them. Some suffered cruelly in the attempt.

One Italian worker at Island Bend had been sending home half his pay packet every fortnight for two years to a girl who had promised to come to Australia and marry him. After his work contract had expired, leaving him free to move to a larger town or city, he made arrangements for the woman to join him. It was then that she wrote to say she had married another man more than a year earlier. The worker went out into the bush one night, sat for a while beside a stream, then shot himself.

Walter Hartwig, the German engineer who had recruited many of the workers, once noted wryly that the Australians used to think it 'bloody marvellous' that there were no women around; but he said it drove the Europeans to despair.

The absence of single women in the mountains was enough to make Dieter Amelung actually consider leaving his much-loved work at the Guthega power station and return to a far less certain future in Germany. At least in Germany he would be able to meet women and find a wife. In Australia it seemed an impossible expectation. For a while he dated the daughter of the postmistress at the Hotel Kosciusko, but the hotel burned down and the girl went off and became a nun. Dieter then took up with the nun's sister, but decided eventually to try his luck in Germany.

While still at Guthega he advertised in a lonely hearts column in a German newspaper and started corresponding with one of the women who replied. He returned to Germany for a holiday and to meet his penfriend. They got on famously, but she did not want to go to Australia.

> I was twenty-seven years of age and very sad. I was in two minds. In Australia I had a good occupation, but my life was going nowhere. In Germany I had met a woman and there were signs that employment there was picking up.
>
> We got engaged and I came back to Australia after she said she would think about it. But a few months later she wrote to say she would not be coming. It was a real blow and I wondered what I would do.
>
> There was very little social life for the single men in the mountains and it didn't look like I would marry here so I began seriously to think about going back to Germany.

Then one day in 1960 when the Scheme was beginning to attract its first tourists and visitors, Dieter was approached in Cabramurra by two women who wanted to know if there was a butcher's shop in town. He told them the nearest was at Kenny's Knob (the little settlement where he had spent his first night).

Neither woman had heard of Kenny's Knob so Dieter offered to drive them. On the spur of the moment he asked if they would like to see the recently finished Tumut-1 power station, where he was now working. They were delighted by the opportunity to visit something they had read so much about in the newspapers and magazines.

Wanting to be friendly to the visitors, Dieter invited them to have a cup of coffee with him.

> They hesitated, but then agreed. Anyway, to cut a long story short, I married one, Heather; and the other became my mother-in-law. They were from near Gosford, north of Sydney, and were holidaying in the Snowy in their new car.
>
> So I married an Australian girl after all, and my English from that day on improved

enormously. It was interesting because up until she met me, Heather, like most Australian girls, had been frightened of migrants, simply because she couldn't understand them.

Despite the thousands of single and lonely men in the mountains, there were no sex crimes, apart from the occasional peeping Tom and underwear snatcher.

One fellow caught stealing ladies' underwear from the staff houses at Cooma North was found to have amassed quite a wardrobe. Nothing, however, was returnable to its owners because he had chewed out the crutches of the panties and the tops of the brassieres. Another chap walked into the Cooma police station to make a tearful confession—wearing the women's underwear he had taken.

One night at Island Bend, the local policeman, Brian Shultz, was on patrol with an Authority patrol officer and special commonwealth constable, 'Taffy' Thomas, when they drove past a man walking along the side of a road carrying a car jack. The two officers stopped their vehicle, thinking the man must need help with a broken-down car further along the road. But when they called to him, he said he was all right. He said he was just out walking and didn't have a car.

The man—a short, stocky Italian—was clearly nervous, and the two constables suspected he had stolen the jack.

'No, is my jack,' he insisted.

'Well, just because it's your jack doesn't explain why you're carrying it around in the middle of the bloody night,' Schultz said.

The two policemen felt something was amiss and decided to take the fellow back to Island Bend for further questioning. He clearly seemed to be up to no good. If he had not stolen the jack then he was out grabbing wheels or something else.

Back in the official atmosphere of the police station the Italian soon broke down and made a tearful confession. Yes, the jack was his and he was not out stealing wheels or anything. He had been out peeping into the women's barracks, but as he was not tall enough to see through the windows, he had taken the jack to jack himself up.

The two officers collapsed in helpless mirth while the little Italian sobbed loudly in mortification. When the constables had control of themselves they sternly confiscated the jack and sent the frustrated Romeo home with a warning to stay away from the women's barracks.

Theft, also, was a rare crime in the construction camps. Although there was a lot of gambling and fighting, men of all nationalities took great pride in being able to leave their paypackets by their beds. If there was a theft they would track down the offender, belt him to within an inch of his life, and then call the police.

Bill Holmes, the Cooma detective, said that in most camps the men policed themselves against theft. It was partly honour, and partly to keep the heat off the illegal gambling, which would be blamed if stealing became too frequent:

> Most were there for a few years to work hard and save money. They didn't want disruptions, so they often sorted out the troublemakers themselves.
>
> It was common to get an anonymous phone call in broken English: 'You look in

room nineteen, hut twenty-seven...you be interested in what you find there.'

So off you'd go to room nineteen and it was odds-on you'd find someone on a wanted list from Sydney or Melbourne. The crims would come to the mountains to lie low, but after a few beers couldn't help talking. The workers didn't want them there and moved to be rid of them as soon as they could.

The only serious case of theft that Holmes had to investigate was when two paymasters employed by the Kaiser consortium had gambled all the money and could not pay the men:

When I went to see them they gave me the key to the safe. It was filled with lottery tickets. They had 2000 pounds left from their gambling binge, so decided to buy lottery tickets to try to win the men's pay back. Not a single ticket won a penny.

When I arrested the two fellows, all one of them could think of was the unfairness of the lottery: 'Two thousand quid...and nothing,' he kept muttering.

The local policemen in the construction communities were often the first bridge between the many different nationalities and cultures.

A policeman was often the only authority from whom a migrant could seek advice or to whom he could take his troubles. Everyone in the vicinity of Cooma was sent to 'Mr Holmes' when they had a problem. As the first detective in the region, 'Mr Holmes' had his own small room at the back of the Cooma police station. There he received a steady stream of visitors, few of whom had any real grasp of English. It made the job burdensome, but also brought its rewards.

One day Bill Holmes was visited by a tall, erect Austrian with a livid duelling scar down one side of his face:

He looked darn near seven foot tall and evil. I wondered what the hell he wanted. From his pocket he pulled a well-thumbed German–English dictionary and word by word indicated he wanted to fill out an application form for his family to immigrate.

I helped him fill out the form; it was simple enough. But then he wanted to pay me. It took a while to convince him that I didn't want any money. Once he accepted that, he stood erect, bowed, clicked his heels and left.

He was one of many, so I soon forgot about him, until one night when I had to go to the Happy Jacks wet canteen with a warrant to arrest someone. As I grabbed the bloke, the rest of the bar moved in around me. They were looking pretty nasty and I wasn't too sure how I was going to get out of it. They had all been drinking heavily and it would be nothing for an anonymous knife to find me.

As they started to press in, shoving and yelling, a thickly accented voice boomed out, 'I on his side.' I looked up, and beside me was the big Austrian, glaring around at the rest of the men, challenging them to defy him. In an instant they changed sides. I suddenly had dozens of smiling, happy escorts showing me to the door with my prisoner.

I will always remember that big Austrian's voice. They were the sweetest words I have ever heard.

CHAPTER TEN

From Jindabyne tunnel and round Island Bend
We boys go to Cooma, our money to spend
And we'll buy youse some beer there, if you happen to see
Four Italians, three Germans, two Yugoslavs and me
We'll pull up in Sharpe Street by the Alpine Hotel
If you've been to Cooma you'll know this place well
And before we get inside our order rings out
Four vinos, three schnappses, two slivovitz, one stout.

From the song 'Cooma Cavaliers' by Ulick O'Boyle and the Settlers.

Jock Wilson, Bev Wales (second from left) and ambulance officer Ron Hunter speak with Lady Delacombe, wife of the Governor of Victoria, Sir Rohan Delacombe after receiving bravery awards for the attempted rescue of two Yugoslav men whose car left the Alpine Way.

THE LAWMEN

If there was one common theme in the lives of all Snowy workers, it was alcohol. It was the elixir for camaraderie and the balm for fear, boredom, homesickness and loneliness.

The wet canteens in the mountains and hotels and nightclubs in Cooma were eventually open twenty-four hours a day to cater for men working the different shifts. Violence and trouble could occur at any hour.

Managing the aftermath of long drinking sessions fell on the shoulders of the handful of policemen scattered through the mountains, and on the Authority's patrol officers. For them, the Snowy was a seven-days-a-week, sixteen-hours-a-day routine from which a man could never relax. The odds against them were sometimes frightening. More than one policeman was transferred out a broken man. In a nation where violent crime was still considered rare and at a time when city streets, let alone the streets of a quiet country town, were safe places for everyone, the Snowy policemen sometimes wondered just where they had been sent.

Cooma gained national notoriety in 1955 after a late night shoot-out in its main street established its reputation as a lawless frontier town. It was a stigma that would last for years.

A girl, described by police as a 'part-time prostitute', was drinking in the upstairs lounge of the Alpine Hotel with a group of Italians after they had all been to see a film. They were laughing and seemed to be having a good time when her boyfriend, Harry Rymal, a Canadian worker on the Scheme, strode in. He had been drinking steadily in another bar. Rymal grabbed at the girl's arm, pointed to the door and yelled at her to get out.

'Let me finish my drink,' she demanded.

Rymal raised his voice to a crescendo: 'Out!' he boomed. The Italians made themselves scarce. They had no desire to be caught in a private argument. Rymal dragged the girl downstairs and out onto the footpath where the two began arguing and yelling at each other.

Suddenly Rymal produced a revolver from within his clothing and started dragging the screaming woman along the street by the hair. They crossed Vale Street and continued on until they were opposite a garage, Balmain's, the site today of the Monaro County Council office. By this time someone had called the police. The 'paddy wagon' arrived, completed a smart U-turn in front of the couple and stopped. Out stepped Senior Constable Bill Graham and Sergeant Fred Chapman, known locally as 'Careless Hands'.

Just as the police alighted from the van, a large crowd began to spill out of the second session at the adjacent movie theatre. The two policemen were approaching the couple when the girl screamed a warning: 'Look out, he's got a gun, he'll shoot the lot of you.' Before the policemen could react, Rymal levelled the pistol at Careless Hands and fired two shots.

To this day no one knows how he missed. Rymal was a recognised marksman and 'Careless' was a big bull of a man, a good axe-handle-and-a-bit across the chest—though he demonstrated surprising alacrity in dancing backwards in his surprise at being shot at.

Rymal then threw the girl on the road, aimed at 'the seat of her employment' and fired a third shot. The bullet inflicted a flesh wound in her upper thigh. She screamed, leapt to her feet in terror and ran to Careless Hands. She fell onto her knees, sobbing, and wrapped her arms around the sergeant's legs.

While this was happening, Bill Graham had dashed back to the van, taken a .32 Webley pistol from the glovebox and fired at Rymal from across the van's bonnet. Moviegoers were flinging themselves behind any scrap of cover they could find. Elsewhere in the town the first shots were beckoning curious townsfolk like a circus call.

But the battle was brief. Bill Graham, the constable generally regarded as the worst shot in the force, fired just three times. Each bullet went into the chest of Harry Rymal, who collapsed onto the road surface.

After the sound of the gunfire the street was suddenly silent except for the sobbing of the wounded girl and a gurgling noise coming from the throat of the slain man. Nobody moved. All eyes were fixed on the formless shadow lying in the dim yellow wash of the street light. The gurgling stopped. Rymal took a deep breath, let it out slowly—and did not take another.

Careless Hands later found one of Rymal's bullets had ripped through the sleeve of his tunic, and the ricochet from the bullet that wounded the girl had taken a chunk from the brickwork of an adjacent store. Where the other bullet went was not discovered until morning, when the garage owner arrived to find petrol still running into the street from a punctured pump hose. All night, unknown to anybody, the main street of Cooma had been a huge petrol bomb just waiting for a spark.

The lot of a Snowy law enforcer was never easy. He often worked alone among hard-drinking, hard-living men, many of whom believed the law was an irrelevance in the mountain frontier. Those who did survive were men who lived by their wits, rather than the rule book. They were tough and had to earn respect, often by proving with their fists that they were better men than those who challenged them. There were some situations, however, that called for a more sensitive response.

Frank Rodwell, a long-serving patrol officer with the Authority, often found peace could be restored after a bloody brawl simply by lending a sympathetic shoulder:

> I learned quickly that some of these fellows had gone through hell and were living on a razor edge. One time I was called to a fight caused by a Polish fellow who was drunk. As I dragged him away he started screaming, 'They're all dead, all dead'. His whole family had been killed in the war and he missed them terribly.
>
> I had another similar experience with a Yugoslav who had opened a mass grave with his bare hands to try to find his mother. Who wouldn't drink and go a bit crazy after something like that?
>
> For me, growing up in Australia and being too young to go to war, it was hard to imagine what they had been through. But you had to learn to understand. You had to let them use your shoulder and cry it all out.

Rodwell did not like to see a man lose his job. He preferred, where possible, to smooth out situations rather than report a man, either to the Authority or the state police.

> Poor bastards, they'd lost everything—and suddenly here's an opportunity to work and earn an honest day's wage. It was paradise for them.

Frank Rodwell, 1965.

The key to keeping control was to develop a good intelligence network. If you learned who was running an illegal gambling ring you would let them know that you knew and keep the operation on a string. That way you could use it as a source of information. A prime suspect for a theft, for example, would be someone who had lost money gambling.

Working alone and unarmed, the patrol officers often found themselves in hostile and dangerous situations, as Frank Rodwell did on the night two big Irishmen started to brawl in the wet canteen at Eucumbene, the camp built for the Eucumbene–Snowy tunnel to Island Bend.

The bar had closed and the men, who were mates, had been working through a tray of beers bought before the bar shut. An argument turned into a fight, which became so fierce everybody else fled as windows, glasses and chairs were smashed. One of the men broke a beer glass on the bar and rammed its jagged edge into the other's face, opening his cheek to the bone.

By this time, Frank Rodwell was on his way, summoned by the barman, Bill Cutting. Frank arrived and looked through the window to size up the situation. He could hardly believe his eyes. The two barrel-chested men, both of whom stood about two metres tall, were still laying into each other. One had lost his shirt and was left with only the remnants of a singlet, and both men were red with blood. There seemed to be blood and broken glass over the entire room.

Frank's only weapon—and often a most effective one—was authoritative bluff. He took a deep breath, marched into the room loudly yelling 'Break it up, break it up' and stood between the staggering combatants. He distracted them by turning their attention to the face of the fellow who had been slashed. It looked as though a small shark had savaged him. 'Christ, we'll have to get that fixed.' He brusquely marched the fellow from the room and drove him to the medical centre.

The nursing sister had spent long enough in construction camps to be unsympathetic to drunken brawlers who called her from bed at ungodly hours. She ordered the blood-streaked hulk into a chair and told Frank to hold the edges of flesh together. 'A big boy like you won't need an anaesthetic, will you?' she said to the wide-eyed Irishman, and proceeded to stitch his face together with a large curved needle. The bloke was tough. He sat through the ordeal clenching his teeth but without a whimper.

It was not a neat sewing job. The Irishman was left with a horrific-looking gash which started under his jaw, curved up near his ear and then down towards his top lip.

Frank drove the man back to his barrack and returned home to bed. He had hardly become comfortable when he was roused by running feet on the gravel driveway and fists pounding on his back door. He opened it to the other Irishman from the fight and his Italian room-mate, whose normally swarthy skin was grey with fright.

They stammered out their story. They had opened the door to the fellow who had been sewn up and he had lunged at them with a big cane-cutting knife. They managed to slam the door on him but moments later the blade splintered through the thin wood and the terrified pair escaped through the window.

Frank told his wife to ring the sergeant at Jindabyne and the Authority's administrative officer at Eucumbene to tell them what had happened and where he was going.

Leaving the two fugitive room-mates at his house, he returned to the darkened camp to search for a crazed Irishman wielding a cane-cutting knife. When he arrived, everybody seemed to be under cover. They had all seen the man, and each time Frank spoke to a different person the size of the knife increased. Rounding the corner of a barrack, Frank bumped into Don Reid, one of the Authority's grounds maintenance workers.

'Seen anybody running around with a big knife?'

'Bloody oath,' Reid said.

'Well, where is he?' Frank asked testily. He was not enthusiastic about wandering blindly in the dark, not knowing where he was going to confront the knife-wielding Irishman.

'Buggered if I know,' said Reid. 'I asked him what he was up to and he belted me fair across the arse with the bloody thing. So I didn't ask any more questions.'

Frank finally found the Irishman in the darkened room of another man with whom he was talking. He turned on the light and held out his hand.

'Right-ho, where's the fucking knife?'

'What knife?'

'The knife you've been carrying around the camp. Hand it over.'

The man went to a locker and withdrew a small cheese knife.

'Cut the crap. You know the one I mean: the big one.'

The man rolled his eyes and shook his head. He denied he had another knife and, as Frank could not find it in the room, he demanded to inspect the Irishman's own room. It was not there either, so he sat him down and asked him to tell the whole story. The man explained that when he got home and looked in the mirror he realised he would be disfigured for life. He had snapped, and had stormed off to seek revenge for this terrible tragedy.

By the time he finished telling his story, the big, rough, tough character was crying—and he could not even remember what had started the argument with his mate. Frank sympathised with him and he calmed down. He assured Frank he had come to his senses and would not cause any more trouble. Frank then tracked down the other Irishman, who had returned and locked himself in his room in a state of shock, and spoke to him.

By about half past two in the morning Frank had the camp calmed down. He returned to his car and, as he had half expected, found the sergeant from Jindabyne and the administrative officer sitting in a Landrover. They had learned from other workers what was going on, and decided they were not going to enter the camp until the rampaging Irishman had been located and subdued.

Frank reported all was now quiet but the waiting pair decided then that they wanted to see the two men who had caused the trouble. Frank urged them to wait till morning; it was always the best way to deal with aggressive drunks.

But they wanted to lay down the law. They went to the men's rooms, hauled them out into the corridor and began to bawl them out. This caused everybody to become worked up again. Frank could see another fight was imminent and he had had enough for one night. Although the sergeant and the administrative officer had seniority, he dragged at them, demanding they let the matter rest until morning when everybody would be calmer. They aquiesced, but reluctantly. This had been *their* big moment.

In the aftermath, the Irishmen left the camp, Frank never found the cane-cutting knife and the sergeant was transferred. The construction camps needed tough but sensible policemen, not men out for an easy bust.

Those who did survive in the tough, knockabout world of the construction camps and towns, and who earned their stripes in the Snowy hall of fame, were men like Fred 'Careless Hands' Chapman, Bill 'Homicide' Holmes, Brian Shultz, Bill 'Chubby' Keefe, Harold Etheridge, Kevin 'Lofty' Lomas, Bob Blissett, Bill Graham, Lionel Whitney, Bill Head, Bruce 'Bags' Durrant, Rod Murray, Bev Wales and 'The Admiral', Ron Davey.

All could handle themselves—Careless Hands and Ron Davey had both been champion wrestlers—and all were colourful characters who fitted into the unconventional way of life in the mountains. Many were regarded as *the* central characters in the communities over which they presided.

During the time Careless Hands was at Cooma, he lived in the house adjacent to the gaol, which was also used as the police lockup. Careless had a passion for gardening and would get prisoners on short-term hard labour to tend his garden. If there were no prisoners he would round up a few drunks from outside the nightclubs, lock them up and let them go sometime the following day after they had put in a few hours work. One day he bought a sack of spring onions which he ordered the prisoners to plant. Careless could never understand why they failed to grow. He seemed to be the only person in the entire town who did not know the prisoners had severed all the bulbs.

The lawmen. Back row, from left: Ron Davey, Arthur 'Ossie' Hann, Harold Etheridge, Bill Holmes, Bill Graham. Front: Bruce Durrant, Lionel Whitney, Fred (Careless Hands) Chapman.

Careless Hands gained his unusual nickname from his fondness for a popular love song of the same name, which he was said to croon in quiet moments. To the locals the hulking policeman—he stood 188 centimetres and weighed 140 kilos—seemed the antithesis of the suave, careless lover in the song.

At Sue City, the workers took great delight in the antics of their local policeman, Ron Davey, and his cherished boxer dog, whose name was Boxer. In the late afternoon just when Ron would be about to taste his first cold beer at the end of a long, arduous day, the valley would echo with Boxer's plaintive wail. To the delight of the regular drinkers, who had come to know the ritual, Ron would storm from the wet canteen, grab a spade and start to climb the steep valley slopes, following the direction of the wailing.

Somewhere up on the mountain, as everybody had come to know, Boxer had tried to chase wombat into its hole. The dog always managed to get stuck about a metre inside the hole and then tell the whole world of its predicament. On locating the animal, Ron would dig it out, slap it furiously across its rump with his trouser belt, then drag it unceremoniously by the collar back to his house and tie it up.

After three or four days Ron would feel sorry for the animal, 'his mate', moping on the end of its leash. In a flood of affection he would let it off for a run. As sure as Monday followed Sunday, on the day the dog was free again Ron would be about to enjoy his first cold ale when the valley would once again echo . . . 'Whooooooo'.

On the nights Ron was forced to clamber over the mountain to rescue Boxer everybody made doubly sure they behaved themselves.

But of all the men who policed the construction camps, only one man became universally regarded as a legend in his own time. He was Constable Bev Wales, the young farm boy from Yass who had tossed his white pith helmet under a Sydney tram in an attempt to evade traffic duties. Bev Wales was renowned as being totally and utterly fearless.

After being given the Cooma posting in 1952, Bev remained in the town until 1955, when he began to nurture the idea of taking over a one-man station up in the mountains. He was about to marry the daughter of one of the Cooma sergeants and did not want to stay too close to home. As well, Cooma work was starting to become routine. There was paperwork and more paperwork, and always to do with the same things—brawls and road accidents:

> The locals and the 'wogs', as they were called, were still feeling each other out. The local lads were toey. They didn't like the young foreigners coming down from the mountains with all their money. The migrant fellows didn't want trouble but they were only prepared to take so much. So there were hundreds of brawls and if a non-Australian was involved, it meant double the paperwork.
>
> Then the fellows would drink all they could buy and try to drive back to camp somewhere up in the mountains. Alcohol, speed, poor roads and extreme weather conditions meant a lot of pretty horrible accidents. Plenty of cars, often loaded with blokes, left the road in places where they dropped hundreds of feet before hitting the ground and exploding.

Bev wanted a beat of his own where he took on all work. In December 1955 he sought, and was happily granted, the job as the replacement constable at Happy Jacks. Happy Jacks had the reputation of being the roughest camp on the Snowy. The workers there had grown accustomed to being a law unto themselves and had virtually driven out the previous constable.

Bev put a lot of thought into how he would approach his new assignment, his first lone command. From the start he decided he would not wear a uniform as his predecessor had done. He saw no point in waving a red flag at rogue bulls. He also realised he had to earn the men's respect using the only language they understood.

To reverse the lawlessness and stay alive, he had to demonstrate beyond doubt that there was no man who could better him. He had to establish that as fact from the moment of the first contact.

On the drive from Cooma to take up his new position he stopped at the wet canteen at Junction Shaft, which was on his new beat and was also where a lot of the Happy Jacks workers drank. There were a dozen or so fellows drinking. Bev stood, framed in the doorway, until he had their attention.

'My name's Bev Wales. I'm the new constable,' he said simply.

For a moment the men looked as though they had not heard him clearly; then they responded with a chorus of whistling and catcalls.

'Just so you all know where I stand, there are laws in this state and I'm going to make sure they're enforced,' he continued. The catcalls got louder. 'So if there's anyone here who thinks he can stop me, then he had better try it now before I unpack my Landrover.'

The wolf whistles and jeers got even louder and four solidly built Australians slid off their stools. It wasn't every day a copper actually asked to have his head punched. Leering, they approached the young constable. They paused just beyond an arm's length, then moved in swinging.

Every account of the incident is the same. Bev Wales used four, maybe five, punches to lay the four men out on the floor. He bundled them into the back of his Landrover and continued on to Happy Jacks to lock them up for the night. Within hours the incident was being recounted throughout the surrounding camps.

The legend of big Bev Wales was only just beginning. Over the next few weeks similar clashes took place as the most aggressive of the workers tried to break the new constable. By the end of his first month Bev Wales had locked up forty men, most of whom spent several nights in a cell nursing bruised jawbones and cut foreheads.

'Over the years they had gone berserk. There was no law and order whatsoever. Living up in the mountains like they were they thought they were beyond the law. The lawlessness had to stop and I was the one sent to do it,' Bev said. 'But after that first month they started to quieten down and for the first time in ages the majority of the townspeople were able to relax.'

Though his beat covered hundreds of square kilometres, he made it his business to know everybody. He talked to people. He introduced himself to every new worker or

Junction Shaft camp where Constable Bev Wales 'introduced' himself to the region's workers. The wet canteen is obscured by buildings in the top of the photograph.

Bev Wales (left) with a mate, Aub Casson, and the police landrover on a road near Happy Jacks.

group of workers who entered his domain. Just by talking with workers he could learn their names and origins. This friendly discourse not only made newcomers aware they had been noticed but gave Bev the basic intelligence data with which to check on backgrounds to learn who might need watching. This way he was usually one step ahead of those looking for trouble.

He rarely worked less than sixteen hours a day. His patrol ranged from Eucumbene Portal near Adaminaby (the entrance to the Eucumbene-Tumut tunnel) to Tolbar Camp, up to Happy Jacks, Cabramurra, Kiandra, Kenny's Knob, Tumut Pond, Tooma camp and all areas in between. He was a lone policeman among 6000 workers and the only links between most of the camps on his beat were rough bulldozer tracks:

> The paperwork was enormous. Whenever these fellows went to Sydney they always seemed to get into trouble—have car accidents, fight and then never turn up to court. My desk would get buried under summonses.
>
> Not only that; whenever the Sydney police couldn't find someone, they immediately decided he'd gone to the Snowy Mountains, so we copped all that work too. A lot of it was chasing up maintenance defaulters—'wife starvers', as we called them. We probably got around to following up about ten per cent of the various summonses that came our way; the rest we just reported as not being there.

Men with a guilty conscience, though, were not to know of the problems faced by overworked policemen. Bill Holmes could not help laughing whenever he drove onto a construction site. At the sight of his car, a dozen or so men could sometimes be seen scattering into the surrounding bush: 'It was obvious they were wanted for something but what they didn't know was that I wouldn't have a clue what it was.'

Tumut Pond wet canteen, June 1953.

Bev Wales backed away from no situation, even when he was outnumbered. When brawling men were told Bev was on the way, his reputation preceded him like a bow wave, often quelling the disturbance before he arrived.

One of the few times he did need assistance was during his first year at Happy Jacks. Six young Yugoslavs had spent a day on slivovitz and under the leadership of a giant of a man nicknamed Big Romeo they ended up holding two or three hundred men under a siege in the Tumut Pond mess.

The ruckus had started with a drunken brawl, during which a supervisor with the Kaiser consortium, Neville Weir, thought if he could get Big Romeo, the rest would settle down. So he drew him into a fight. It was a mistake from the start and Weir was already just about all-in when another Yugoslav thrust a broken glass into his face, cutting it to ribbons. The cowardly attack inflamed the mob and the fight turned into a brawl involving hundreds of men.

The six Yugoslavs who started it rushed outside and started hurling tunnel rock through the windows, showering slivers of glass over everyone. It stopped the brawl and sent men scrambling for cover; but as the bar and most of the tables were smashed, there was not much shelter to be had.

The whole camp was trapped in the mess, everyone too scared to poke his head out of the door for risk of being brained by a piece of tunnel rock. Urgent medical help was also needed for Weir. Luckily there was a phone in the mess and a call was put through to Bev Wales, thirty kilometres away at Happy Jacks.

There were no other policemen in the area, so Bev phoned his brother, Rod, who was also at Happy Jacks and asked him to bring a mate of his, Brian Clempson. They were both big fellows who could handle themselves. While he was waiting for them to arrive at his house, the phone rang again: 'They've just left in an old Zephyr. They heard you were coming so Big Romeo reckons he'll get you first. They've taken the old back track.'

When his two offsiders arrived, Bev put a Webley pistol in his own pocket, gave his brother a baton and Brian some handcuffs. Thus armed, they headed off along the track to Tumut Pond. Five kilometres from the dam at the Tumut Pond crossing they saw the Yugoslavs' car parked on the opposite bank. It was now dark and Bev trained the Landrover's lights on it. At the first sign of violence, he told the two men with him, 'Get straight into them, because if Big Romeo gets you first you probably won't get up again.'

With the Landrover's lights on the car Bev could see every move the Yugoslavs made. Big Romeo started rolling up his sleeves, obviously meaning business. Bev led the way across the bridge, and as he approached Big Romeo began winding up for a punch.

The policeman hit first: 'I barrelled him, got a good one in first and down he went. He was so shocked he thought I'd shot him. We then got all the others out of their car and handcuffed them. They had no fight once Big Romeo had gone down.'

He handcuffed Big Romeo to the glass-slasher, a fellow named Balde, and jammed the whole six into the back of the Landrover. Bev told Rod to drive while he and Brian kept an eye on their prisoners: 'We started back to Happy Jacks. But Big Romeo recovered and tried to get at us. It was pretty nasty. They were desperate, and I could tell they would do whatever was necessary to get away—so I had to thump him a few more times to settle him down.' At Happy Jacks, Bev charged the men and called Cooma for a paddy wagon. He didn't trust the Happy Jacks lockup to hold the men.

Big Romeo and the assailant with the broken glass each received gaol sentences. Some years later when Bev Wales was stationed in Khancoban, after the Scheme had shifted from the Tumut development to the Murray development, he was suddenly confronted in the street by the big Yugoslav, who invited him to have a drink: 'You could see straight away he was a different man. Gaol had done him some good. So we shook hands and had a beer.'

Bev Wales's most unnerving encounter, and the only time he ever consciously confronted his own mortality, occurred in 1957 at Kenny's Knob.

The call came one evening from the camp industrial officer, Gordon Bird, one of the few English-speaking people there: 'Got a bit of a crisis, Bev. One of the blokes has stolen a revolver from the night watchman and is shooting the lights out in the canteen. He's had a few drinks, so some of the blokes are getting a bit edgy about it.'

Bev cursed under his breath. It was midwinter and freezing outside. He put the Webley into his deep overcoat pocket and drove to Kenny's Knob—an hour and a half away in the near-blizzard conditions. When he arrived, Gordon Bird apologised for having dragged him out into the night: 'It's all over,' he said. 'Once he'd fired off all the rounds he gave the pistol back and has gone to bed.'

Bev said it was not over until he had seen the man. It was a serious offence and he intended to lock him up. Gordon Bird gave him the directions to the man's barrack and Bev knocked on the door.

'It's the police. Open the door, I want to see you.'

There was a pause, then a thickly accented voice called, 'Go away. I shoot you.' Bev assumed he was bluffing. He already knew he no longer had the pistol.

'Come on. Open the door or I'll kick it open.'

'No. I shoot you.'

After being called out into the freezing cold Bev was in no mood to play games. He kicked in the door and pulled up short with a double-barrelled shotgun pointed straight at his stomach. Because he thought the man had been bluffing he had not taken the pistol from his pocket and he now had no chance of drawing it.

'I going to shoot you,' the man said, his finger curling around the trigger.

Bev cursed himself for not having his hand on his gun or in his pocket because he would not have hesitated to shoot through the coat. All he could do now was try to keep the man's mind off the trigger: 'I kept my eyes on his face and could tell he wasn't rational. He was a slightly built fellow in his early-to-mid-thirties but his eyes looked like those of an old man whose mind had gone.'

Bev also noticed that the gun, still pointing at him, had three barrels—two twelve-gauge shotgun barrels and a .32 calibre rifle barrel. Both shotgun hammers were cocked. He remembered Gordon Bird telling him the man was a 'gun freak'.

'I've never seen a gun like that before. I'm interested in guns,' he said. It was a masterly understatement. At that moment he was very interested in that particular gun. 'What sort is it?' he asked.

The man began to tell him. Bev, who had his eyes glued to the man's face, saw a flicker of distraction and in one swift movement grabbed the barrels and whipped the butt around to club the man across the side of the head. The fellow began to struggle but, as he was slightly built, Bev easily subdued him. When he opened the gun Bev found it fully loaded.

Bev took the man to a doctor to have the lacerations caused by the blow with the gun butt treated and also to have the doctor sign a schedule under the Lunacy Act. The man had clearly suffered a mental breakdown. Bev drove through the night to deliver him to a psychiatric hospital at Goulburn, 300 kilometres away.

When he got back to Happy Jacks about midday the following day, the Goulburn hospital telephoned. They had just found the man dead in his cell. He had hanged himself. Bev later learned the man was Hungarian and had had a history of instability since the war. He had never recovered from whatever tragedies he had suffered and had gradually gone to pieces.

Bev spent sixteen years in the mountains while the Snowy was under construction. He survived through a mixture of police fieldcraft and bush psychology shaped by the instincts of a boy from the bush. He attributes his success in handling the men to his decision not to wear a uniform and not to carry a baton: 'Bring out a baton in a brawl and you would put everyone offside. But take a man on with just your fists and the others respected you. If you did get into trouble they'd even help you; but belt one of their mates with a baton and they'd kill you.'

What most endeared Bev Wales to law-abider and offender alike was that he obviously cared for people. Like Frank Rodwell, the Snowy patrol officer, he often let bush justice take its course, rather than invoke the kind of officialdom that might lose a man his job:

Often I'd arrest a bloke and he'd say, 'Let me fight you, Mr Wales. If I win you let me go, eh?'

The fellows were afraid of losing their jobs, so I'd say okay. We'd pair up in the lockup, exchange a few blows till I gave him a good thump and it was over.

It was summary justice, something you couldn't do today, but it kept the peace there. If you beat a man in a fair fight it also allowed him to save face and keep his dignity among his mates.

They had no respect for anybody who was soft. They were tough fellows and they expected you to be the same. You had to meet them on equal terms.

Despite these 'wild west' elements, the major preoccupation for Wales, and for all policemen in the mountains, was the horrific death toll, both on the roads and at the construction sites:

At Khancoban alone I handled ninety inquests, and no death was clean. Men were cut in half, decapitated, blown to pieces, or squashed by huge boulders.

Death was frequent and it was part of my job. Tunnels collapsed, air hoses burst, truck brakes failed and bulldozers and trucks and cars toppled into ravines.

You had to handle a lot of grief and tragedy. The road fatalities were often the hardest for people to accept when men who had survived the dangerous work were killed doing something that should have been safe and routine.

Like one of the Kaiser bosses, Whitey Giesel, whose Landrover skidded on ice and toppled into Rainbow Creek near Happy Jacks. The vehicle landed upside-down on its hood in just two feet of water, but both he and his wife drowned. They had only recently been married. She was Australian and had been so excited about going to America to live. It took us an hour to get them out, and all the time the motor was running. They were people we all knew really well, which always made it very hard. When we laid their bodies on the ground about an inch of ice formed around them, it was so cold.

In 1963 Bev Wales, an ambulance driver, Ron Hunter and the indefatigable Scot, Jock Wilson, were awarded bravery medals by the Royal Humane Society of Australasia for the attempted rescue of two men—Tomislav Kraljevic and Mirko Kulis—whose vehicle had failed to take a bend on the Alpine Way and had hurtled through space to the bottom of a gorge.

The award came after the Victorian Ambulance Service had recommended that the participants in the operation be acknowledged and rewarded. It actually got Bev into trouble because he had not considered the event to be out of the ordinary and had, therefore, not reported it to any higher command.

At the accident scene there was no way of ascertaining whether there were any survivors.

The three men had to climb about a hundred metres in darkness down a near-vertical cliff to reach the victims who, sadly, were dead.

During his time at Happy Jacks, Bev Wales also helped solve one of Australia's most intriguing and enduring aviation mysteries.

On Sunday, 26 October 1958, Thomas Sonter, a young carpenter working on the Tooma Dam project with Thiess Brothers, set out with his camera to take photographs of the scenery in a remote part of the Toolong Mountains. As he was clambering through the undergrowth on a steep slope above a tributary of the Tooma River, he almost tripped over a structure of rusted tubular steel lying among the trees. Sonter recognised the framework as that of a plane and returned to camp with as many of the small metal fittings as he could carry.

He told fellow workers back at the Deep Creek canteen of his find but refused to reveal the location. Sonter wanted the reward, should there be bullion or diamonds in the wreck. The other workers refused to believe him so he produced the plane's brass identification plate. Another worker, Ken Bladen, made a note of the numbers and set off to find Bev Wales. Sonter, meanwhile, packed his bags and left for Cooma to catch a plane to Sydney and stake his claim.

With Bladen's piece of paper in front of him, Bev Wales telephoned the Department of Civil Aviation in Sydney. He told an officer that a plane had been found in the mountains and read out the serial numbers. The officer said he would ring back. Some minutes passed before the phone rang. The man was breathless with excitement: 'If what you have quoted me is correct, it's the *Southern Cloud*. It's been missing for nearly thirty years.'

Even Bev was excited. Most Australians knew about the mystery of the disappearance of the Avro-10 plane *Southern Cloud* on a flight from Sydney to Melbourne on 21 March 1931. Its loss had sounded the death knell for the new company, Australian National Airways, founded by pioneer aviators Kingsford Smith and Charles Ulm to provide the first regular service between Australian capital cities, and no trace of the aircraft, its crew and six passengers had ever been found.

Bev immediately phoned Bill Holmes in Cooma, telling him to grab Sonter at the airport and bring him back to Happy Jacks where civil aviation inspectors would be arriving the next day. Sonter was told there was no reward, just a lot of trouble for himself if he refused to show them the plane's location.

News of the find flashed around the world and Bev spent the remainder of the day answering questions from journalists and press agencies from every country he had ever heard of. The first journalists began arriving that very day, despatched at a moment's notice by news editors in Sydney and Melbourne. By the next morning, when the aviation inspectors arrived, there were at least fifty journalists, photographers and cine cameramen among the entourage, as well as a number of senior police detectives from Sydney.

The group, led by Sonter, trekked up through thick scrub from the proposed site of the Deep Creek dam to a peak known as World's End Mountain. There, among the snow gums, lay the haunting, crumpled remains of the *Southern Cloud*.

The crash site was 350 kilometres from Sydney, thirty kilometres east of the pilot's intended route—not bad considering he had flown unexpectedly into wild 160-kilometre-

Thomas Sonter holds a magneto as he stands with one foot on
the centre motor of the wrecked plane.
PHOTO: THE AGE

an-hour winds—and just below 1400 metres above sea level.

The forward and centre sections of the fuselage were badly telescoped. The centre and starboard engines were buried a metre into the ground in correct alignment to each other, but the port engine was more than a metre forward of the centre engine, indicating the plane was banking sharply to starboard just before impact. Fragments of molten metal indicated the aircraft caught fire on, or shortly after, impact. As a result of bush fires and foraging animals over the intervening years, the only human remains were a few small pieces of bone and the bones of a foot in a shoe. Other relics found in the

The *Southern Cloud*.
PHOTO: THE AGE

surrounding earth included two wrist watches. One was too badly burnt to be of use to the investigators, but the other had its hands stuck on 1315 hours—a quarter past one in the afternoon.

The most remarkable find was made by Bev Wales—a Stanley thermos flask. When he carefully unscrewed the lid, the coffee it contained still had a strong, fresh aroma. It was a tangible link with the people who had been on the plane during their last terrifying moments:

> I don't know why, but it suddenly meant something very special to me so I put it carefully aside.
> Then one of the detectives from Sydney came along and put a pick through it to see what was in it. It nearly broke my heart. After all those years, and after surviving the initial impact, some moron comes and does that. I couldn't believe anybody could be so thoughtless and stupid.

It was probably the only time anyone had seen the young constable, so renowned for his toughness and fearlessness, come close to crying.

Workers salvaging wreckage at the Southern Cloud crash site.
PHOTO: THE AGE

CHAPTER ELEVEN

Happy Jacks school, June 1956. Kalev Tarmo is shaping the snowman, with bare hands. Today no trace exists of what for a while was Australia's highest town. Kalev now works as a mechanical engineer with the Snowy Mountains Authority.

WINTER

WINTER

As a boy growing up in the wilderness around the small, isolated township of Happy Jacks life for Kalev Tarmo was one big adventure. He was a typical Australian bush kid—exploring, fishing, chasing wombats and going to school in a tiny one-classroom school.

Yet only five years earlier, in 1951, Kalev had been sailing into a frightening unknown on an Australia-bound refugee ship with his parents, who had decided they could never return to their homeland, Estonia. In the refugee camp they had left he had scavenged for cigarette butts, removing the shreds of tobacco until he had a full tin. His father then bartered the tobacco for food.

Now it was only a memory. He could scarcely recall what life before Happy Jacks had been like. He was a little Aussie. All the children in the school were—except that they spoke Dutch, German, French, Norwegian, Swedish and half a dozen other languages. They also wore different clothes—the Austrians wore leather shorts, the Norwegians thick woollen breeches, and the Americans wore denim jeans. The Europeans were united as young new Australians by the challenge of having to learn a common new language. They were living, learning and playing together without regard to national origins.

The isolation in the bush welded them, instilling in them a sense of self-reliance and equality. By the time they shifted with their families to other parts of Australia they would know more of the secrets of the Australian wilderness than native-born Australians raised in the towns and cities; and be able to tell, with authority, about the great scheme being built in the Snowy Mountains. Living on the roof of the continent, in itself beyond the experience of most Australians, the children became acutely attuned to the seasons. Free of the distractions of adulthood they knew, to the day, the turning point from one season to the next:

One morning you would go outside and there would be water on the ground. Later that same day you would see some birds and you knew the winter was over and that spring was coming. In a fortnight or so you would know the skies would clear and that you would awake to bright, crisp mornings and the singing of hundreds of birds. It meant we could start exploring the bush again as the snow melted and the first wildflowers appeared. Even as kids we knew it was beautiful.

The expectation of summer was one of the strongest emotions guiding their lives. Summer was freedom. The winters, on the other hand, were confining. They spent weeks at a time shut inside their homes or their classroom. Often the little school would be buried in snow and the children had to enter and leave through a snow tunnel. It was fun for a while but the novelty wore off after weeks of being hunched behind cold wooden desks with chattering teeth and runny noses.

One of the few entertainments to look forward to in winter was wombat riding. After each new snowfall, young boys would wait close behind the known location of a wombat burrow. Before long the wombat would emerge to clear its burrow of snow. The intending jockey grabbed an animal and straddled it for a crazy, careering race down the snow-covered hillside. The most dangerous part of the ride was at the finish when boy and cranky wombat had to part company without the wombat souveniring shreds of breeches and skin.

The winters in the primitive conditions of the Snowy camps and towns imposed a harsh existence. Bulldozer and grader operators worked round the clock clearing snow to try to keep the roads open. Sometimes it was an impossible task and the Authority required every household to keep at least three weeks supply of canned food in the cupboards for times when communities were cut off.

Yet little attempt was made to modify the style of housing for the conditions. Prefabricated weatherboard cottages, which would have served perfectly well in central Australia, were simply transplanted into the mountains. Each house even had a typical suburban Hills Hoist clothesline. Housewives would often open their curtains in the morning to see frozen washing sitting on top of two metres of snow. To retrieve the clothes meant an hour or more of digging snow away from the hoist.

Winter created endless domestic problems for housewives and young mothers. Ironically, considering the basis of the Scheme, the most precious household resource in winter was water. Pipes often froze and burst, leaving a house without water for washing or cooking. Mothers with babies had a terrible time.

Sheets had to be cut up for use as nappies and when these ran out nappies which had only been wet were simply hung up to dry before being reused: 'The smell was enough to make your eyes water,' recalled Guiditta Fabbro. 'Nappies and sick babies and no water made life very hard. I only had one baby, but some women had two or

Cabramurra, winter 1955.

WINTER

The outskirts of Happy Jacks, winter 1956.

Aftermath of a blizzard—iced-up weather station Windy Creek, 1959.

three. And they were always getting wet in the snow—not that it was much drier inside. Water condensed on the inside of the walls in the day and froze at night so you would wake up every morning surrounded by ice.'

When warned of bad weather the women in the mountain construction towns soon learned to fill baths for emergency washing water. For cooking it became almost the norm to trudge outside, fill buckets of snow, and sit them on the stove to melt. As soon as the road was cleared, some took their laundry to Cooma—but only 'staff' wives could afford to do this.

For workers, the winters were no mere inconvenience; they represented yet another life-threatening danger. Work and progress were expected to be maintained, regardless of the weather. As a result, iced-over roads, white-outs and the risk of being lost and stranded all exacerbated the potentially lethal working conditions. The Authority, which had to pay contractors 1000 pounds a day when roads were closed, pushed road-clearing teams mercilessly.

Snow-clearing units operated as a convoy—bulldozer, grader, a Landrover with men and shovels to dig out the grader if the need arose, and another Landrover with provisions. Once out on a road there was no chance of returning to the mess for a hot meal. For plant operators the job was hell. The graders had doorless cabs with flimsy canvas curtains as the only protection against the sheeting snow and ice. The front windows had no wipers; they were designed instead to open to provide 'clear' vision.

The plight of the bulldozer drivers, however, was far worse. They sat hunched over their levers, exposed to the full force of the elements. It was not uncommon for operators

Snow-clearing on the Kiandra–Tumut Pond road, 1953.

Kiandra in winter, August 1956.

to be forced to work up to thirty-two hours without a break. They laboured on treacherous mountain roads with weak lighting and sometimes—day or night—they could barely see the front of their machines.

Blizzards were a constant threat to workers toiling in the capricious mountain climate, as one team of drillers who were preparing the foundations for Tantangara Dam, at the head of the Murrumbidgee, learned in the winter of 1955.

It was mid-July and had been a mild winter so far; but one Friday afternoon Max Paterick, who had grown up in the area, warned the camp foreman there was a bad storm on the way. He said he was going to try to get home to Kiandra before it hit. Max at this stage was back in the transport division after his brief fling as a tunneller, and was working at the site as a packhorse handler, ferrying men and provisions down the steep track to the base of the proposed dam.

Normally Max spent his weekends at the camp, but his senses were telling him to get out. He was going to take one of the horses but the foreman, who had looked at the grey sky and decided it seemed placid enough, offered him a lift in his jeep, since he was going to Cabramurra for the weekend. Max accepted the ride and threw his pack and skis into the back. Several kilometres from the camp they crossed a creek. Max advised the foreman to stop while he jammed a tall stick into the ground to mark the crossing. 'You never know. If we get a heavy snow you wouldn't know where the bridge was,' he explained to his puzzled driver.

A little further on, as they slithered their way through the sludge of the last snowfall, the vehicle bogged to the chassis. As it was too far to walk back to Tantangara camp, Max donned his skis to cut across the top of the range to Kiandra. From there he could send word to Cabramurra for someone to pick up the stranded foreman.

The foreman was fortunate that Max knew his way about the mountains and was not long in reaching home to make the call. Shortly after he had been collected and delivered safely to Cabramurra, the storm struck with terrifying ferocity.

The grey sky blackened and the mountains were plunged into a howling, freezing white-out which brought all outdoor work to a stop. Men huddled in barracks and huts, playing cards and restlessly waiting out the storm. Efforts to keep the roads open were in vain. The storm was unremitting, and after five days the twenty-five men sheltering at Tantangara ran out of food. An attempt was made to deliver supplies but the vehicles could not get through.

Late on the sixth night, the phone ran in the Patericks' sleeping household. The caller was Bill Hudson. The commissioner told Max about the men still stranded at Tantangara. He said that one of them had left the camp earlier that day to shoot rabbits for food and had not returned. He asked if Max could get them out. Max was practical and blunt: 'Not tonight I can't. There are three or four flooded creeks between here and there. But I can have a go first thing in the morning.' Even in daylight visibility was minimal; at night it would have been impossible.

As the grey dawn light filtered through the raging blizzard, the phone rang again. It was Hudson, saying the camp had radioed to inform Cooma headquarters that the missing man had found his way back. Max remained unimpressed: 'Well you'd still better make some arrangements to get them out of there,' he told the commissioner. 'I don't know how long this storm will last and someone's going to get killed if they keep wandering about for food or if they try to walk out.' Hudson told the young man he would despatch two bulldozers to open the road to Kiandra and send a vehicle to collect Max. It would take him to Adaminaby from where a rescue effort could be launched.

It was mid-morning before the road was cleared, and midday before Max reached the spot where it began its descent to Adaminaby. Rather than go all the way to the town, Max decided to set out alone from this point on skis. Time was crucial. He did not want to be out in the storm after nightfall, with men whose physical and mental condition may have been precarious. He told the driver to stop, saying he would ski across the mountains to the camp and use the packhorses he had left stabled to bring out the men. He asked for bulldozers to follow as far as they could to enable trucks to meet them somewhere on their outward leg.

It took more than two hours for Max to cover the twenty kilometres to the camp. When he arrived he found the men hungry but in good spirits. They had been living off parrots and rabbits they had shot. Max wasted no time. He explained what they were going to do and saddled the horses while the men dressed themselves as warmly as possible.

He sat the men two to a horse and led from the front on skis. The workers, initially, were keen to leave the camp and return to where they could enjoy a proper meal, but their enthusiasm died the moment they left the lee of the mountain. As they climbed onto the plain they were struck with the full force of the blizzard—a 160-kilometre-an-hour wind, filled with sheeting ice and snow.

The men and the horses immediately wanted to return to the comparative safety of the camp. It took all Max's powers of persuasion to convince the men they had to keep going; but Max's biggest fear was the horses. The further they progressed, the more nervous they became. Their whinnying and tossing made the men more frightened.

By the time they came to the first creek crossing Max was worried. He was one man against twenty-five and he knew that if they rebelled and tried to find their way back to the camp they would get lost and perish in a matter of hours. Most of them were Italians and Australians. They had neither the clothing nor the experience to survive a night in a blizzard.

Max stopped near where he knew the creek ran. He strained his eyes, trying to descry some detail in the swirling white-out. The creek was well concealed beneath the snow but he could just make out the stick he had planted to mark the bridge. But he could not remember whether he had placed the marker on the right or left side of the bridge. He thought he had marked the right edge, which meant they would need to keep to the left of the stick. But if he was wrong, they would plunge through the snow into the near-frozen water. He rubbed his forehead with his frosty gloves, trying to think clearly. If he did not find the bridge, they would all die; they had travelled too far to turn back to the camp.

Max moved forward and began to probe the snow with his ski stocks, feeling either for the bank of the creek or the wooden planks of the bridge. As he moved cautiously about, one of the men slid from his horse, trudged through the deep snow and impatiently shook Max's shoulder: 'I find it,' he yelled. His words were snatched away by the wind, and before Max could react, the man plunged forward through the snow, and dropped from sight. He had found the creek.

Max glided quickly to the edge of the hole. The man's head was about thirty centimetres below the level of the snow and he was standing waist-deep in the frozen creek. He was already babbling incoherently, clearly shocked and rapidly freezing. Max was able to pull him out by lying flat on the snow and reaching under his shoulders. He dragged him straight to the leading horse and, with cord from his pack, tied the man's hands to the horse's tail. Knowing now where the bridge was, Max led the party across and up to a clump of trees on a rise. Tied to the tail of the horse, the fellow who had fallen into the creek was forced to walk and keep his blood circulating.

In the shelter of the trees, Max used kerosene and charcoal he carried in his pack to make a small fire. He spent fifteen minutes warming the frozen man as best he could, especially his hands and feet which he feared would be frostbitten. But it was growing dark and Max knew they had to keep moving or they would all freeze. He again tied the man's hands to the tail of a horse and urged the dispirited procession into movement.

They battled on against the drifting snow for another eight kilometres, Max ceaselessly prodding the frozen man tied to the horse, forcing him to keep walking. Time itself seemed to freeze as the group moved slowly and painfully across the storm-lashed plain. After what seemed an eternity they saw in the distance the lights of a bulldozer and, behind it, the yellow glow of vehicles which were edging forward.

When contact was made with the rescue party, Max handed over his charges, who were fed hot food and drink and trucked down to Adaminaby. Still there was no respite for the high-country horseman; it would be hours before he could shelter from the cold. His priority now was the horses. He had to lead them on through the blizzard to Adaminaby and then find them feed and a sheltered paddock.

The worker who fell into the creek was taken to hospital with frostbite. He lost a toe but was fortunate not to lose his life.

In the winter of 1967 a prolonged blizzard also trapped twelve men at Junction Shaft. By this time, however, snowcats—small cabin vehicles on four pontoon caterpillar tracks—had replaced horses. Len Neithe from the transport division was sent out from Cabramurra with an electrician, Gert Rymere, and a young labourer to bring out the men.

The three men headed out into a white-out. Snowdrifts covered the landscape, obliterating landmarks. They followed the line of the roadway as best they could but, while crossing one particularly high drift, the vehicle tipped to its left. It was just enough for the battery leads to swing loose, touch the steering column and short out the electrics. They were stranded.

It was about eight o'clock at night and Len radioed for someone to bring out another battery. He gave their estimated location, assuming they had kept reasonably close to the road, and the three men huddled together to wait. It was a very cold night, well below freezing, and the men were inside what was basically just a steel cabin fitted with thin canvas seats. They kept rubbing their ankles and stamping their feet to keep some circulation in them, and moved about in the cabin as much as they could.

Before long it was clear to the two older men that they had a problem with the young labourer, who kept falling asleep. They took it in turns to shake and slap him to force him to stay awake. Len knew that if any one of them fell asleep death would claim him

A snowcat similar to the one in which Len Neithe was
stranded and later drove across a frozen dam.

by morning:

> The young bloke was really scared. He was in his early twenties. It was his first time out in the snow and he was convinced he was going to die. It took a lot of talking to convince him we'd all be okay as long as we stayed awake.
>
> Gert Rymere kept telling him about Germany and the war—anything to distract him from sleep and keep his mind alert. It was hard because we all wanted to go to sleep. Snow is dangerous and unforgiving and lulls you into thinking you will feel warm by sleeping.
>
> We were wearing Authority issue greasy wool socks, ski boots, warm shirts and parkas and were still in agony from the cold. If you spat on the floor it was ice before the spittle hit; you daren't let any skin touch the steel cabin or you would have had to cut it off.
>
> The thing that worried us most was the fact that as we had tipped down an embankment it meant we weren't on the road. So we didn't really know where we were, or whether anyone had heard our radio call.

All the men had to eat in the snowcat were emergency rations—two small tins of sardines and salmon and two packets of Sao biscuits. The cold snapped the key for the tins on the first turn, so they were reduced to smoking cigarettes and eating the biscuits.

The men were stranded in their frozen steel cell until three o'clock the following afternoon. They heard a cooee, and a hydrologist skied up to them with a battery strapped

Guthega Dam under winter ice.

Winter 1956. The staff quarters at Island Bend.

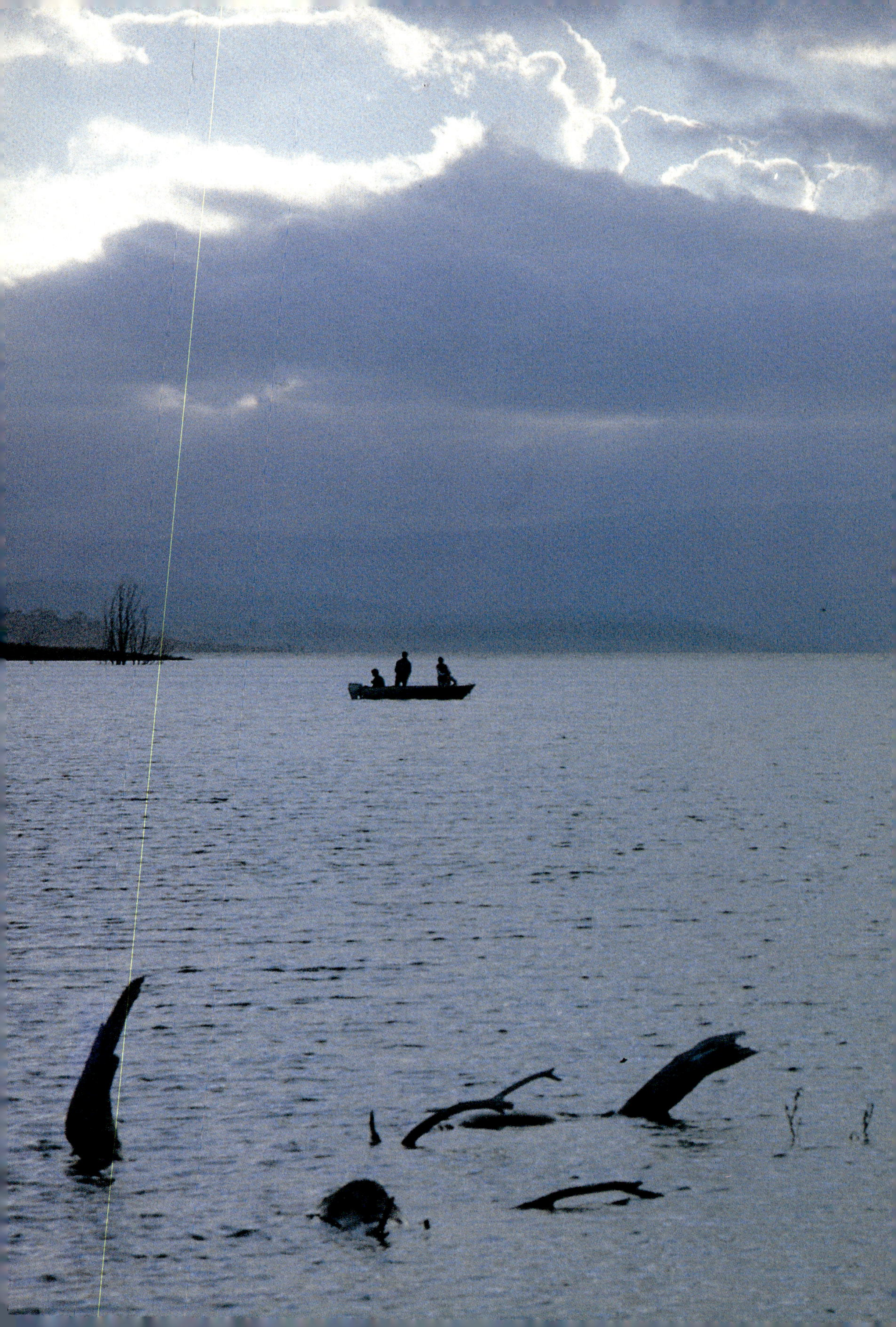

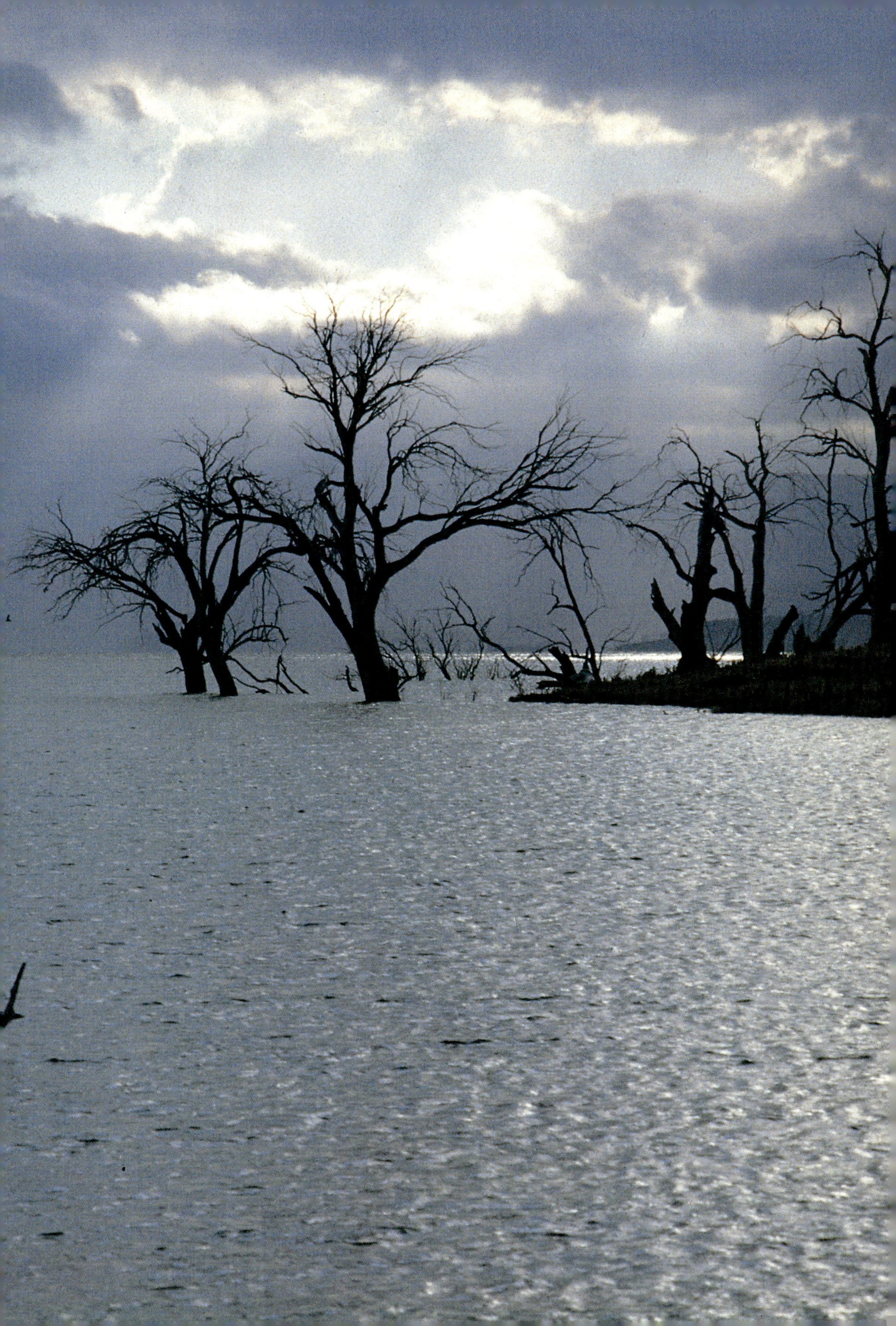

Jindabyne before the deluge.

Previous pages
Lake Eucumbene—an inland sea where once sheep grazed and stockmen drove.

to his back. The men were grateful to be found at last, but Len was not impressed with the time it had taken. Back at Cabramurra people denied that anyone had heard their radio call. If that was so, Len asked, why was a hydrologist skiing the Alps with a snowcat battery strapped to his back? 'Basically, no one else had wanted to risk going out in the blizzard. We could have died out there,' he said, disgusted.

As it was, his heels turned black from frostbite and he later had to have a large part of them cut away: 'At the time it was part of the job. But looking back I often wonder how we survived.'

The next winter he very nearly did not survive. One night a power failure plunged the Cabramurra township into darkness. Len Neithe was despatched in a snowcat to take three electricians to Tolbar switching station, where it was assumed the overload switch had tripped the fuse. It was a 32-kilometre trip from Cabramurra to Kiandra over Sawyers Hill. Len managed to deliver the electricians to the switching station by midnight. The job was finished quickly and the group headed back to Cabramurra. On the way, Len decided to take a short cut along the old Kings Cross road.

As the snow had obscured most landmarks Len kept the transmission towers on his right as a guide. They were just visible in the reflected wash from his headlights. The snowcat sped through the night and Len kept his eyes on the sporadic outline of the towers. He knew that if he strayed too far left he would run into Dry Dam, which had been rebuilt and now supplied Cabramurra with its water supply.

Len Neithe (far R) with mates at Cabramurra in 1963.

The men arrived back safely and Len went to his room, where he tunnelled gratefully under blankets and went to sleep.

The next morning, after just a few hours rest, Len was awakened by the voice of the transport supervisor which bellowed through the depot, 'Who was the bloody idiot who drove across Dry Dam last night?' Len ignored the man. No one could drive across the dam. The ice would not be thick enough to take a vehicle.

Len clocked on for work and was talking to a fellow who had just come back from a weather station: 'Someone did, you know,' the man said.

'Did what?'

'Drive across the dam—and you're the only other bloke who's been out that way since yesterday.'

Len studied the man's face and felt his stomach tighten. He donned his thick snow jacket and trudged across the open ground to take a look, following the incoming tracks left in the early hours by his snowcat: 'Sure enough, there were my tracks right across the middle. I felt sick. It gave me a hell of a fright. I must either have been going too fast to break through or someone above was looking after us.'

In the snow, all the dams were potential deathtraps for the unwary. A number of workers from northern Europe made the mistake of thinking they could skate on the ice. Most were lucky and survived crashing through, but two were drowned while attempting to skate on Three Mile Dam.

Although the winter often brought hardship, drama and tragedy, the presence of European workers on the winter slopes also awakened Australia at large to the season's pleasures and the prospect of a new sport—skiing.

Until the Scheme began to push roads through the mountains, recreational skiing in New South Wales was limited to the gentle slopes near Kiandra and a few slopes below Mount Kosciusko. The Kiandra chalet, the Hotel Kosciusko at Digger's Creek (near the subsequent turn-off to Island Bend) and a government-owned chalet at Charlotte's Pass near Mount Kosciusko were the only resorts. The slopes used were natural; there was no such thing as chairlifts, and skiing itself was considered the realm of the more adventurous of the Sydney and Melbourne social sets.

Apart from those who worked or lived in the mountains, few Australians gave the sport a second thought. The first changes came with the arrival of 400 Norwegians, who were brought out by Selmer Engineering for the Guthega project.

The pipeline bench that had been cleared for the massive aqueduct which would channel water into the power station's turbines was too great a temptation for the Norwegians after the first winter snows blanketed the area. They built a jump at the bottom and turned the steep bench into a devil-may-care slope to test even the hardiest nerves. The sight of one of these expert skiers, hurtling down the steep slope above the power station and flying high over the heads of onlookers like a giant, graceful bird, was memorable for all who witnessed it.

In 1953 the Norwegians organised a ski carnival on the slopes of Perisher Mountain as part of their celebration of the birthday of King Haakon of Norway. They were just having fun, doing something they had learned to do from the moment they could walk; but their activity captured people's imagination and inspired magazine and sports writers

Guthega power station, 1956, showing the pipeline bench down which the Norwegian workers first thrilled the locals with their dare-devil skiing.

around the country. In less than two decades Perisher Mountain would become synonymous with Australian skiing.

The Norwegians, however, were not interested in developing the sport in Australia, as most of them intended returning home as soon as their contract was finished. The initial impetus for the sport's development in the Snowy Mountains came from a single man, Tony Sponar, a champion Czech skier who fled his country after the Communist takeover.

Tony represented Czechoslovakia in the 1948 Olympics and during that year and the following year was the Czech national champion. As such he was allowed to travel with the team to other European countries to compete and train. In 1949 while he was in Austria, he defected, along with the Czech women's champion, Sasha Nekvapil. Tony had intended going either to the United States or Canada because of the skiing there, but those countries had five- and ten-year waiting lists for immigrants.

In Austria, he had married an English girl who worked for the Australian Immigration Commission, attached to the International Refugee Organisation. It was suggested to him that he think about going to Australia: 'It has snow; it just needs to be developed,' he was told. He applied for a job as a winter ski instructor at Charlotte's Pass, the only resort in Australia which offered organised instruction.

But his hopes of lifting the sport in his new country were almost dashed from the beginning. A week after he and his wife arrived the Hotel Kosciusko burned down, severely curtailing the limited accommodation facilities in the mountains. It was a hard blow just when interest in the sport seemed to be picking up.

Skiing's future looked uncertain and Tony was bored as an instructor. While considering

Tony Sponar on his way to fifth place in the 1948 Olympics.

his options he contracted tuberculosis, and the matter was taken from his hands. He was told by doctors to find less strenuous work. As there was nothing obvious in the offing he decided to ignore his ailment and accept a job with a hydrology section on the Snowy Authority. Most of the section were Czechs and as he was their national ski champion he had little trouble getting the right introductions.

The hydrologists travelled on skis for all their winter surveying and Tony thought the job would be a breeze. Unfortunately the first thing asked was, 'Do you ride a horse?' Tony replied, 'In a fashion.' Even that was overstating his ability in a saddle. 'Okay,' came the response. 'Get ready for a week out in the bush.'

The team spent a week on horseback, looking for a site to erect a gauging station on the upper reaches of the Murray River. While moving about the mountains with the hydrology team, Tony kept a constant watch for long, challenging ski runs on which he could establish his own resort. He still believed skiing had a future in the Snowy and was keen to be involved in its development.

There were no maps, and his only way of determining possible sites was by exploration and listening to the locals. Over several winters he skied once a month from Spencer's Creek to a gauging station at Dead Horse Gap and in doing so became familiar with the Thredbo Valley. It was there, above the Crackenback River, that he found the slopes he had been seeking. As well as having the right mix of gradients, the runs also had the advantage of facing south-east. This meant they would be partially shaded from the fierce Australian sun which, even at that altitude, could ruin a good snow base.

During this period Tony learned the Authority was planning to build a road through the valley to skirt the highest ranges and link up with the Murray side of the Scheme. He started talking about the potential of skiing to anybody in the Authority who would listen.

Meanwhile, Sasha Nekvapil had also migrated to Australia and had replaced him as the ski instructor at Charlotte's Pass. Most Australians had been dismissing Tony's ideas for European-style ski resorts as nonsense, so Sasha introduced Tony to a Sydney architect, Eric Nichols. Nichols listened to his plans and began to promote them. As a publicity exercise to increase awareness of the Snowy's potential for skiing, Nichols invited a Swiss chairlift manufacturer to Australia.

By 1956 Tony had selected his sites for a resort at Thredbo. But he could do little more than look at them and dream; they were totally inaccessible. However, one morning Tony's superior called him to his office and told him that Sir William Hudson wanted to see him. It was unusual for the commissioner to want to see a mere worker, and Tony wondered what he might have done to warrant such personal attention.

Hudson did not waste words: 'I've heard about your ideas for a ski resort in the Thredbo Valley. You might be interested to know we're going to put a road through there. If at any time you feel we can be of help, we will do what we can. We can't spend public money on a private project, but since we'll have equipment there...'

Tony got the message. He was in Hudson's office for less than ten minutes and emerged on top of the world. Hudson remarked wryly to an engineer, 'The new Australians are becoming like old ones and the old ones are becoming like new ones.'

The following year Tony Sponar was granted permission by the New South Wales government to develop the slopes, and he left the Authority to devote his energies to creating what would become one of the most popular winter destinations in Australia. It was a risk to leave a good job when he had a family to consider, but he had great faith in the sport at which he had so excelled. He built two huts and installed a primitive rope tow on the first Thredbo slope. The same year he formed a syndicate to get the proposal off the ground, and in 1958 launched a public company. Later that year the Thredbo Lodge was opened and Australia's first chairlift was installed.

Shortly after the lodge opened, Tony had a disagreement with the rest of the syndicate and left. He gained the lease to the burnt-out Hotel Kosciusko, which was an eyesore, and rebuilt it as Sponar's Lakeside Inn. At the same time he established another nearby slope. He also became instrumental as a consultant for the development of Mount Buffalo in Victoria and Mount Blue Cow.

It was not long before the seeds of his pioneering efforts began to bear fruit and Australians started to flock to the slopes, awakened to skiing's thrills by the immigrant workers and given access to the slopes by roads built for the construction of the Scheme. The New South Wales National Parks and Wildlife Service today estimates that more than a million people a year visit the ski fields.

CHAPTER TWELVE

Boyd Mould, 1987.
PHOTO: VERN HUNT

FLOODS AND REBELLION

By 1959–60, ten years into the Scheme, the mountains were reverberating to the thunder of heavy machinery. The Snowy Mountains Scheme was not only becoming a reality, it had gained an unstoppable momentum, carrying on its back the pride of an entire nation. For the first time in Australia's history the world was giving it due recognition as an industrial power in the making.

The engineering expertise deployed on the Snowy was in the vanguard of construction technology. The Scheme was training engineers and other professionals who for the next thirty years would head the world's biggest hydro-electric and construction projects.

The first trans-mountain diversion of water through the Eucumbene–Tumut tunnel to the Tumut River was made in June 1959. By 1960 the lower Tumut development was almost finished. All that awaited completion were the Tumut Dam, the Tumut-2 power station, the Murrumbidgee–Eucumbene tunnel and the Tooma–Tumut tunnel. At this stage the Guthega power station had been feeding electricity into the New South Wales grid for five years. The Tumut Pond, Eucumbene, Guthega, Happy Jacks and Tantangara dams were finished, as was the massive Eucumbene–Tumut tunnel and the Tumut-1 underground power station. It had been a phenomenal effort which had also by now secured the solid backing of the federal government and the approval of the states.

The Authority was now preparing for the third group of contracts—the Eucumbene–Snowy tunnel, the Snowy–Geehi tunnel and Island Bend Dam. This was the start of the second half of the Scheme, the Snowy–Murray development—though the teams of surveyors and geologists had long since left the Tumut regions to lay the foundations for the Murray projects.

The contracts for the first Murray projects were announced in 1960 and Prime Minister Menzies, gearing for an election, issued a statement that work would create 4000 jobs. He neglected to mention that the workforce had already been gathered and that a large number of other workers were soon to be retrenched as Tumut projects were completed.

Men looking for work, as well as dismissed labourers from the Tumut projects, flooded into Cooma. By the end of spring about 4000 men were in the town, most of them homeless and penniless migrants. The Anglican rector of Cooma, the Reverend Frank Woodwell, opened up the church hall as a soup kitchen and emergency hostel. Many men, however, were still forced to sleep in surrounding farm sheds and under bridges. Woodwell alerted the government and the Snowy Authority to the wretched condition to which the political sleight of hand had reduced thousands of innocent men; but he was rebuffed callously.

In an attempt to discredit Woodwell's claims, the Authority's patrol officers were ordered to visit the hall at nights to take headcounts which could be used to disprove the existence of a small army of unemployed. Prime Minister Menzies summoned Hudson to a meeting in Canberra to discuss the matter. On his return to Cooma, Hudson called the mayor, Alderman Johnson, and Woodwell to his office and told them the local police inspector had officially denied the reports and that, therefore, the 'unemployment issue was dead'.

Woodwell, however, had already checked the records of the local police station, and these totally discredited the alleged statement by the police inspector. The two men wrote a formal letter pointing this out to Hudson. They then called a public meeting. The local community responded generously, with a local unnamed SP bookmaker offering

Start of work on the Tumut Pond Dam, March 1957.

FLOODS AND REBELLION

Tumut Pond Dam under construction, January 1958.

The awesome wall of the Tumut Pond Dam nearing completion, May 1958.

The concrete-arch Tumut Pond Dam at completion, 1959. The dam collects inflow from the Tumut River plus water diverted through the Tooma–Tumut and Eucumbene–Tumut tunnels to provide the headwater for the Tumut-1 power station.

free hot meals to every unemployed worker for two weeks.

Severely chastened by what had become a shameful blemish on the Authority's reputation, Hudson despatched a junior officer to call discreetly on Woodwell and offer whatever stores, blankets and equipment were needed.

The following year the Menzies government announced the successful tenderers for the first Murray projects, and the second part of the Snowy Scheme began. But it was cold comfort for the thousands of men who had been lured to the Scheme by dishonest politicking.

The main contractors for the Eucumbene–Snowy and Snowy–Geehi tunnels and Island Bend Dam were a joint venture comprising Utah Australia–Brown and Root Sudamericana, and Thiess Brothers.

The 23.5-kilometre Eucumbene–Snowy tunnel would be the link between the Murray and Tumut developments. It would take water from the Snowy River at the Island Bend pondage for storage in Lake Eucumbene and would return it when needed for diversion through the fifteen-kilometre Snowy–Geehi tunnel to the Murray power stations, and ultimately the Murray River. Later a tunnel would link Island Bend with the Jindabyne Reservoir (Lake Jindabyne). This would provide the Scheme with two vast, interlinked water reserves at Eucumbene and Jindabyne.

Before any of this could occur two entire townships had to be shifted to make way for the rising waters of the two huge man-made lakes. The first to go was Adaminaby, a small town serving the grazing industry, though around the turn of the century it had also been a copper mining centre. A nearby settlement, Kyloe, produced 4000 tonnes of copper between 1872 and 1913. The mine and its 213-metre shaft were now to vanish under an inland sea.

The valley of the Eucumbene River at Adaminaby was the key to the entire Scheme. It was the main collection point for water flowing off the eastern slopes. A giant dam, one of the highest in the world, was built across the Eucumbene River thirteen kilometres below the township. The dam checked the waters of the river to form a storage in the river gorge. Once that was filled it would spill out to fill the vast Eucumbene valley and low-lying areas of the surrounding plains.

The water did not creep out from the gorge and begin to spread across the valley until mid-1957, eighteen months after the dam wall was finished. By this time, the last of the reluctant landholders had been moved out. A new township had been constructed eight kilometres away on a grazing property, Bolairo View, but many residents of the old town elected to leave the district.

There had been long and often heated argument over just where the new town should be sited. Many argued in favour of simply moving higher up the slope above where the existing town lay, but proponents of the Bolairo site argued that the large body of water would enshroud such a site for long periods in cold fog. They believed that the Bolairo site, being in the lee of the mountains, would be milder and less prone to snow. 'We'll be two jumpers better off,' said the town fathers who were pushing for the Bolairo site. Another factor said to favour Bolairo was that it would be astride the main connecting highway that linked the Monaro with the western districts on the other side of the mountains.

A poll was taken and the Bolairo site won out. The Authority employed a town planner, Professor Denis Winston, to lay out the new town. It also planted thousands of trees and shrubs on the otherwise treeless plain.

Despite efforts to improve the aesthetics of the site, the chosen location proved in hindsight to be a bad decision. The distance from the lake robbed the town of the huge tourist developments which later took place at Jindabyne where similarly displaced residents opted for a lakeside settlement.

Adaminaby residents were given the choice of either selling their houses to the Authority, at market value plus ten per cent, or having them transported to the new town. Most of those who intended remaining in the district had their homes shifted in preference to selling and seeing generations of memories bulldozed.

In all, 101 buildings, including two churches, were moved from the old township. New buildings constructed at the Bolairo site included a modern shopping centre, hotel, hall, theatre, post office, police station, two schools, church buildings and two service stations.

When the publican of the old town's hotel decided to leave, the Authority was left holding both the new hotel and its licence. It ran the hotel in the new town for two years until a number of federal parliamentarians complained that it was improper for

The first houses begin moving out from Old Adaminaby, 1956.

a government authority to be operating a pub. The Authority was forced to sell at a loss in order to appease its political masters with a speedy sale.

At the time the first houses were being moved, the water backing up from the dam had yet to appear, and some refused to believe it would ever reach the town. These sceptics maintained their position to the end. Others, though, were quite keen to move; the new town would have electricity and sewerage and would mean an end to kerosene lanterns and outside 'dunnies'.

The main opposition to the flooding of the valley came from the graziers whose farms would be inundated. Many were the fourth generation of their families in the district and were aggrieved by the loss of both their land and their heritage. One grazier, Morris Mackay, collected a sizeable sum of money to fight the Scheme in court but in the end did not proceed with the action. Few people had the spirit to oppose what had been inevitable for years.

Two of those affected by the flooding of their land were Dolly and Dudley Bolton who, as newlyweds, had only just started farming their own property, which included the site of the dam wall itself. Dolly was a fourth generation Crowe. Her great-grandfather, Tom Crowe, had taken up land in the valley in the 1830s and her parents, Bill and Freda, still ran the original property on the plain just outside Adaminaby.

It took two years for the water to rise to within sight of the house in which Dolly had grown up. It was a daunting sight, watching familiar paddocks become an inland sea, but the family had little time to linger and admire their new view. Landholders in the flood path had been warned that once the water was on the plain it would rise rapidly. As the waters neared the house they drove before them hundreds of snakes which took refuge in the building. This forced Dolly's mother into tearful recognition that the end had finally come.

Dolly's parents departed on a Saturday morning when the water was almost lapping at the steps. They had to cut a new track through the bush, as the road to the farm was already immersed. The following afternoon, when the family returned to collect a copper washing tub, they found the house had already sunk from sight.

The historic homesteads on the properties bought by the Authority were demolished and the materials sold to unaffected landholders for use in building sheds and other farm buildings. In this way entire houses were sold in pieces for sums of between ten shillings and twenty pounds.

The removal of the town and the displacement of graziers virtually decimated the local community. Many people, including most of the uprooted farmers, left the district. After the racecourse and showgrounds were moved to their new home they fell into disuse through lack of interest.

Dolly and Dudley Bolton went to live in Cooma for two years, but were able to return and resume farming when they bought a property at Middlingbank, on the southernmost arm of the new lake.

When they returned in 1959 the site of their original home was under eighty lightless metres of water. As for their town, the lower end of the main street lay hidden beneath about seventy metres of water, charcoal-grey under a brooding sky. At the top of the hill the old school, which Dolly had attended as a child, remained as a solitary sentinel over the town's watery gravesite. Today, a person looking towards the lake from the school, which still stands, gazes out over the side streets that branched off to the right of the old main street.

In 1973 the rising waters of Lake Eucumbene reached their highest mark. Once the Scheme became fully operational they stabilised at a level about twenty metres lower. The drop in the water level in 1974 allowed the recovery of the lake's first drowning victims. One skeleton, still wearing fisherman's gumboots, was removed from the dead branches of a previously submerged tree. Another skeleton, with a concrete block chained to a bony ankle, was found on the muddy bank. Its discovery solved a local murder-suicide mystery. A year or so earlier a woman had been found drowned in a farm dam, along with a note from her husband confessing to murder and stating that he intended killing himself. Police suspicions that he may simply have left the area and gone into hiding were dispelled by forensic tests on the skeleton. To date, twelve people have drowned in Lake Eucumbene. Several, including three Victorian policemen who went missing on a fishing trip, have never been found.

The flooding was a traumatic time for the people of the Monaro but there was one unexpected and unusual episode to distract them. In July 1959 rumours that a major film was to be shot in the area were confirmed with the news that international stars Peter Ustinov, Robert Mitchum and Deborah Kerr were seeking local houses to rent. It soon transpired they were coming to the district to film *The Sundowners*, a drama about outback life, based on the book by the Australian writer Jon Cleary.

The eager anticipation became actuality three months later when 140 cast and crew moved in as temporary members of the Monaro community. Robert Mitchum raised eyebrows by driving himself down from Sydney, while technicians and crew from Warner Brothers arrived on a charter flight direct from London. Most of the filming was carried out in the small township of Nimmitabel, which had to have its only strip of bitumen—its main street—regravelled for the film.

The people were proud their district had been chosen for such an international event. But shortly after filming began it rained solidly for a week. The local population was inconsolable, and everybody from the postman to the shire council apologised profusely for the bad weather. When Warner Brothers' executives announced they would have to move the shoot to South Australia after heavy rain washed out the big bushfire sequence a feeling of gloom settled over the district. The company executives were both astonished and warmed by the way the community had so passionately involved itself in the production, so they decided to stay put and hold out for dry weather. They rewarded the community for its support by inviting everybody to become the crowd extras in the country race meeting scene at the resited Adaminaby racecourse.

The district was buoyed by the decision. The atmosphere became festive but the cast and crew of the film were openly saddened when they learned the beautiful town and valley of nearby Jindabyne were to be flooded.

Jindabyne was a small bush town serving the local grazing community. It nestled along the banks of the Snowy River in the Jindabyne valley, renowned for its tranquil beauty. Where the Snowy flowed through Jindabyne it was broad and gentle, flanked

Jindabyne on the banks of the Snowy River, 1953.

by poplars and willows. It was a popular haunt of one of Australia's greatest poets, Andrew 'Banjo' Paterson.

The town had grown along the banks of the Snowy River at a crossing on the cattle route from the Monaro plains to Victoria. In 1893 a sturdy wooden bridge replaced the ford crossing and the settlement grew into a little town. By the 1950s there were two general stores with low verandahs, a row of unpainted cottages and a post office. Higher up on the hill was a Catholic church. Nearby on another hill was the Church of England, with briar roses clawing over its walls. There was also a public school and teacher's residence.

The houses which lined the dirt road to the bridge were simple hand-built weatherboard cottages, with cast-iron kitchen stoves which in many cases provided the only heating during the bitter winter months. The churches were built of brick and stone and designed to last till Judgement Day—which for Jindabyne was delivered by the hand of man, and a mite sooner than the town's forefathers had anticipated.

When the Snowy Scheme started in 1949, Jindabyne's population was about two hundred and fifty. Almost all the people of the Jindabyne valley were descendants of the pioneers who had settled the district in the 1830s. They had a reputation as stoic battlers but surrendered without fuss to the powers who wanted their valley. When it came, the order to shift was no surprise. The local people had always believed the Snowy River had an important destiny to fulfil.

They accepted that their town at the foot of the mountains had become a strategic point in the government's plans. The flooded valley would create the main water storage for the Murray-1 and Murray-2 power stations and the Murray valley irrigation areas in northern Victoria and south-central New South Wales. The town had already been the base for the road building projects—first Guthega and Island Bend, and later the Alpine Way which established a road link through the Thredbo valley and around the bottom of the Great Dividing Range to the Murray development.

The Authority's first camp in the town had been three kilometres downstream from the bridge at the site of an old flour mill which had been built to grist wheat grown in the district before people decided to concentrate on grazing.

The people of Jindabyne were given eight years notice of the Authority's intention to dam the river and flood the valley—eight years in which to watch their town slowly die. Townsfolk would gather in the hotel and talk about what the Scheme was doing to their lives. All new building in the town had been stopped and even the cemetery closed. While not openly opposing the Scheme, many saw the possibility of a reprieve in the political uncertainty that faced the Scheme; but in 1958, with the commonwealth-states agreement, even that flicker of hope died.

Arrangements to relocate Jindabyne began in 1959. After its experience with Adaminaby, the Authority decided against giving the Jindabyne residents a choice of new sites. It arbitrarily selected a site on the valley's upper slopes, which would put the new town on the shores of the lake formed by the damming of the river.

The relocation of Jindabyne was achieved in a much more positive spirit than was the experience at Adaminaby. The Authority took more care to involve the people in the design of their new town and worked hard to foster a perception that something better was being built.

Creating an instant town was a mammoth task. Roads, sewerage, power, drainage, water supplies, schools, shopping centre—all had to be planned, and new access roads built for farming properties cut off by the water. Most Jindabyne people approved of the new site and welcomed the many improvements offered by modern 'city' houses equipped with the latest household conveniences. The Authority planted trees to green and shade what was a naturally bare hill, and gardening competitions were organised to hasten the town's establishment and to develop a sense of community spirit.

The relocation was completed in 1967 and the townsfolk said farewell to the old town with a pageant and symbolic final crossing of the big, old wooden bridge which had been the town's landmark since the turn of the century. The residents, some driving bullocks and dressed in period costume, filed up out of the valley to their new home on the hill.

Behind them, an army demolition team moved in, using the bridge as a structure on which to test its skills. Snowy workers looking on were astonished at the antiquated methods and materials used in the demolition. The army team placed sticks of gelignite and then produced a type of fuse that civilian engineers had not seen for decades. According to Alf Harvey, a senior officer with the Authority's property division: 'It seemed pretty silly. With all the modern plastic explosives and expertise available on the Snowy, the army moved in with something from the last century. Still, it brought the bridge down well enough.'

As the boom of the explosion reverberated through the valley, many of the townspeople cried. When the smoke cleared, the bridge, the final link with their heritage, was gone. Two momentarily stunned platypuses drew a loud collective sigh from onlookers as they floated listlessly before reviving and swimming away.

Later, when the water began to fill the valley, Alf Harvey watched as the first farm his great-great-grandfather owned gradually was submerged by the rising, dark tide. But memories of the old town soon dimmed. They were further diminished during the first summer of a full lake when the townspeople, who had lived most of their lives well away from the coast, discovered the attractions of water sports on their new inland sea.

Before the removal and demolition of Jindabyne the Authority undertook a systematic collation of historical records and items of interest from the town. Organisations such as the National Library had reacted with horror at the way no such effort was made at Adaminaby. Many of Adaminaby's historic cottages were simply bulldozed, with no thought given to salvaging items of cultural value.

The cost of moving Jindabyne was $2.8 million, but $1 million was recouped from the sale of land as the new townsite attracted tourism interests and expanded. Removal, which mainly involved rebuilding, was completed in 1967 and within three years the town's population doubled. Today the population is still growing as Jindabyne rides a tourism boom.

One of the few distresses suffered by the Jindabyne people, which the Adaminaby people were spared, was the exhumation of human remains from the cemetery and their removal to a new site away from the water. The Adaminaby cemetery was high on a hill, beyond the reach of the encroaching waters; but the little Jindabyne cemetery was inundated.

The government considered leaving the cemetery alone, but was advised that it would be exposed at times of low water level. There was concern that wave action might erode the ground over the graves and litter the shoreline with the bones of the town's former citizens. The government cleared the way for relocation of the cemetery with a special Act of Parliament and gazetted a new site; but there was a problem: most of the graves were unmarked and it had been eight years since the last burial. It was considered necessary to identify the individual graves. Living in a small close-knit community, the Jindabyne people had had no qualms in the past about leaving the identities of the graves' occupants to the memory of the living. Besides, the nearest stonemason was in Sydney.

The Authority advertised nationally its intention to move the cemetery and set a date, 10 December 1961, for a gathering at the graveyard to allow relatives to locate and identify gravesites. The Authority located eighty-four graves in the cemetery, of which fifty-two had no headstones. It intended to rectify this in the new cemetery by providing a stonemason and ordering relatives to arrange for headstones to be put over the new graves. Where there were no relatives to pay for a headstone, the Authority would meet the cost. The exhumation and reinterment of the remains from the old Jindabyne cemetery was carried out by a Sydney contract gravedigger and stonemason named Luscombe.

Property officer Alf Harvey was placed in charge of the cemetery removal because he had been on a burial party during the war. It was a dubious qualification but was sufficient for his superiors, especially as nobody particularly wanted the job.

Alf was present at the gathering organised to identify the graves at Jindabyne: 'There were some interesting arguments as people pondered over just where Grandma or Uncle Bill had been buried. Once we got the general location, we determined the precise spot with a crowbar. The earth over a grave, even after fifty years, was still softer than the surrounding unturned earth.'

When the process of marking all the graves had been completed, it turned out that there were more graves than recorded burials. It was decided to attribute the discrepancy to shoddy book-keeping rather than launch a tedious and time-consuming investigation.

The method of exhumation was to place the remains in pillowcases, which were then placed in new caskets. 'But only four-foot-six coffins, because there's not much left after fifty years,' explained Alf.

If there were no remains we put a shovelful of the 'sacred earth' into the pillow case.

Some of the remains were preserved better than others. In most graves there was nothing but a few bones and the handles of the coffins. One bloke we dug up had been buried in a specially-made redwood coffin secured with four-inch copper nails. It was the most complete skeleton I had ever seen. The sand, soil and air hadn't got to him at all.

He, of course, needed a full-length coffin. So did another fellow who had been embalmed. I heard the digger at the bottom of the grave muttering about needing a pick to get this one in a pillow case. I looked down and just about died. There was a fellow as good as the day he was buried, so we carefully transferred him also to a new full-size coffin.

An interesting discovery made by the team was that, contrary to the law of the day which said bodies brought back for burial from interstate or overseas had to be in lead-lined coffins, they found only galvanised iron in such graves. Some undertaker in the past had no doubt increased his profit by the substitution, never imagining it would ever be discovered.

Contrary to what some had expected, Alf claimed the diggers found no jewellery. 'I used to get ribbed in the pub: "How many gold teeth did you get today?" But it was nonsense. All we ever found was Masonic regalia in a few of the graves.'

It was unpleasant work and, even though it was carried out in midwinter, the smell was sickening. The bodies in the galvanised iron coffins had taken a long time to rot, while other graves had suffered from water seepage. Even when a body had completely decomposed, the soil had absorbed the smell. When Alf got home each night his wife barred him from the house until he had stripped naked and hung his clothes on a line to air.

The relocation of the cemetery was completed in July 1962 and, on return to normal duties, Alf's workmates presented him with an undertaker's hat. His role in the cemetery removal, however, was not easily forgotten. His car had acquired a rather putrid odour, and the Authority eventually had to replace the vehicle.

Although the townspeople of Jindabyne and Adaminaby allowed themselves to be uprooted and transplanted with a minimum of protest, there were some who refused to go without a fight. Boyd Mould, the sixteen-year-old who had shared the railway carriage with garlic-breathing migrants on his return home from holidays, was now, more than a decade later, trying to establish his own property south-east of Jindabyne near Berridale.

He deeply resented the changes imposed on the people of the Monaro and in particular the phasing out of High Plains stock leases. He had shared the community's apprehension when word first spread of the Authority's plans for big dams and the flooding of some of the old families' homes and land.

With the final abolition of alpine leases in 1969, resentment became open hostility. Boyd was attempting to fulfil his boyhood ambition and pay off his own farm, which meant he needed to run as many stock as possible. To do this the home paddocks had to be rested, but both the commonwealth and the New South Wales governments had agreed the mountains should be closed to stock. Overstocking had already caused serious erosion and the quality of water was vital to the future of the Scheme. The Commonwealth had calculated the leases were now worth seventy dollars an acre from their contribution to power generation and irrigation, compared with the ten cents an acre earned from the grazing fees.

The graziers moved out but with reluctance and anger. In the summer of 1967–68 drought struck the Monaro and sheep were starving on the scorched plain. In January the wind started to blow and did not stop until May. Entire fencelines disappeared under sand-drifts. The position was desperate. The wool market was depressed, so the graziers could not sell their sheep. The choices were limited to shooting the animals or taking them back into the mountains.

Boyd was one of many who decided to 'damn them all' and take his stock back to where he knew there was abundant undergrowth choking the alpine bush. Retracing the old stock routes he drove a big flock up onto the family's old High Plains lease. Regardless of what the authorities did to him, he at least knew his sheep would be spared the agony of starvation.

His relief, however, was shortlived. A week later he heard a vehicle pass his gate and through the plume of dust saw a fully laden stock truck. He knew, instinctively, it was his sheep on the way to the pound. He waited for more trucks to pass, but there were none. His hopes rose. Only one truck meant the rest of the flock was still in the mountains. When Boyd returned to the old lease in the autumn, more than 2000 of his flock had survived.

The drought extended into the following summer and Boyd and several other graziers again took stock back to the mountains. It was like old times—the droving, campfires, yarns, bush tucker and healthy stock. The qualities they believed had made Australians resilient and great were living on, and they were imbued with confidence that they could beat the bureaucrats who were trying to snuff them out.

When the summer of 1969–70 brought still no relief from the searing drought, the men of the Monaro once more drove their stock up into high country. They put an estimated 100 000 sheep and 15 000 head of cattle back on the old alpine leases and the National Parks and Wildlife Service decided it was time to crush the practice once and for all.

The graziers were alerted by bush telegraph that the 'National Parks' was planning a major sweep. The rangers mounted a massive muster, loading sheep into every truck they could lay their hands on from Yass to Cooma, as well as their own truck, which could carry ninety sheep. The sheep were taken to old stockyards at the foot of the mountains.

Boyd was convinced the rangers were primarily out to catch him and another grazier named Bob McMillan:

They brought down about 3000 sheep and yarded them on Snowy Plain. The rangers were pretty pleased with themselves, but the next day there were about eighteen graziers at the yards checking out the tags. When they were sorted there were none of mine and only about 180 of Bob McMillan's.

The rangers camped by the yards, intending to take the sheep the next day to Bombala; but in the morning the sheep were gone, right from under their noses.

Anyway the buggers reckoned I did it. I didn't, but they purposely went back into the mountains especially to look for my sheep, identifiable by a dye brand comprising an arrow with a dot above it. They rounded up about 2000 and brought them down to Botheram Plain, near the Gungahlin River.

Few of the rangers had experience in stock handling. To make the flock more manageable, they intended separating the sheep into smaller mobs and trucking them to holding yards scattered throughout the region—Bombala, Rocky Plains Reserve and even as far as Queanbeyan. By this stage, however, no local transport operators would carry the sheep and the rangers had to wait for an operator from Yass.

Boyd was able to monitor their plans through the bush telegraph. He learned his sheep were to be trucked out the following morning. Not a man to surrender to officialdom without a fight, he left home at midnight, arriving in the vicinity of the holding yards at about one o'clock. He parked his truck about three kilometres away, intending to release the sheep quietly and herd them to where the truck was parked. It was a warm night with a full moon to work by. He felt that with his best dog, which he knew would work quietly, it would be an easy operation.

Leaving the truck, he walked across the dry grassy plain to the perimeter of the camp. After a quick reconnaissance he realised that whoever was in charge was not stupid. The sliprails that formed the entrance to the yards faced the river. Instead of being able to slip the rails and move the sheep straight out onto the plain, he would have to herd them in a sharp turn away from the river and past the rangers' camp on the opposite bank.

Still, what had to be done, had to be done. He slipped the rails and, working quietly with the dog, freed the sheep and began moving them around the yards and campsite. All seemed to be going well, until one of the last sheep in the pens took fright and rushed the gate. That triggered a noisy stampede, made all the more obvious and ludicrous by the loud tingling of bells. Boyd had forgotten about the bells put on the animals to help trace them during musters in the mountains.

A light appeared immediately in one of the nearby tents. Throwing caution to the wind, Boyd began hustling the flock as fast as he and the dog could manage. Listening to the voices on the bank, Boyd judged there to be only two men and a woman guarding the sheep, so he felt he still had a good chance of getting them away.

However, the ranger in charge of the operation was not easily duped—he was Bob Leech, but no longer the fresh-faced eighteen-year-old he had been in the early days of Island Bend. After abandoning his 'education' among the former stormtroopers, Bob had graduated as an agronomist. He worked with the New South Wales Department of Agriculture for five years, then joined the National Parks and Wildlife Service.

It was Bob who had been put in charge of the campaign to sweep the graziers and their stock from the mountains. His side of the story is that he rounded up Boyd Mould's sheep because they were the closest to the loading area he had set up. He had not suspected Boyd at all over the disappearance of the first flock to be impounded. Nor had he been asleep when Boyd turned up, because earlier that day another grazier whose sheep had been impounded had threatened to turn up with an armed gang to forcibly take back the sheep. Consequently, his forces were larger than Boyd had calculated. He had with him four other rangers, his wife Helen and her Samoyed dog, named Snowy:

> The moment we realised someone was trying to move the sheep out, we jumped from our swags, saddled the horses and crossed the river. We could see a shadowy figure moving the stock away so we went around to cut him off.
>
> We held him for a while but then our dog, Snowy, decided to change sides and help Boyd's dog; so he started getting away again.

In the dull moonlight the sheep were a swirling mass of grey, moving in an erratic stampede away from the pens. Boyd was bemused, but pleased, by the defection of the 'Husky' which was happily following the lead given by his sheepdog. The flock see-sawed for about an hour as the opposing sides fought for dominance, but eventually it was Boyd who slowly gained the upper hand. His confidence grew as he realised he was making progress. He had not spoken so knew there was a good chance he had not been identified.

Despite the illumination offered by the moon, his droving direction was upset by the pursuers, and the sheep suddenly baulked when they came up against a thin, scrubby creek. The lost momentum gave Bob Leech and his rangers the chance to regroup in front of the mob at the creek and Boyd, realising he was beaten, surrendered.

'I thought you'd have more sense, Boydie,' said Bob when he recognised the grazier. Boyd shrugged, his attention distracted by another ranger who was chasing his dog with a large stick. He called the animal to his side and left the rangers to put the sheep back in the yards.

Boyd was not sure what to do next, but was determined to have his sheep back. He was in debt because of his efforts to develop his own farm, and he could ill afford to lose them. He returned home, prepared sandwiches and a flask of tea, went back to the holding yards and camped in the loading ramp. In the morning, when the rangers tried to load the sheep, Boyd refused to move, and Bob Leech radioed his headquarters at Sawpit Creek, asking for someone to send the police.

The police evidently were not all that interested in becoming involved. It took eight hours for a sergeant from Jindabyne to arrive, and by that time it was too late in the day to move the sheep. Boyd knew the sergeant well; he had a son in the local football team Boyd coached.

'I'm being victimised,' Boyd told the sergeant.

'Maybe, but the law is the law, Boydie. You can stay and keep an eye on your sheep but don't do anything foolish,' he warned.

Boyd camped nearby and early next morning two policemen from Cooma arrived. Boyd told his story again. They were sympathetic but explained that if he tried to prevent the sheep being moved they would have to arrest him. The following day the sheep were loaded aboard trucks and were taken to Boyd's father's property because there was no pound big enough to take them all. In the process, ewes and new lambs were separated, which further fuelled the grazing community's general resentment towards the rangers and their government masters.

Boyd eventually retrieved his sheep for an impounding fee of four cents a head. From Bob Leech's viewpoint, he had done a pretty cheap muster for the grazier. The following year the long drought broke and the desperate need to find alternative pastures eased.

In time Boyd Mould came to agree with the decision not to allow grazing in the higher plains. It was traumatic at the time for the graziers, but people adjusted and, with modern pasture improvement, were able to lift stocking rates on the home properties. The bush community was not decimated, as some had feared. Boyd points out that the names on yesterday's headstones, in today's telephone directories, and on kindergarten rolls through the Monaro remain the same.

However, the grazing community still refuses to accept its total prohibition from the mountains. They point with cynical detachment to the 'ravaging' by a burgeoning ski industry and to the ungrazed, unfired forest floor and warn again of the type of holocaust that devastated tens of thousands of hectares in 1985. The graziers also complain about their stock being savaged by marauding packs of wild dogs which raid from their haven in the national park.

Boyd Mould says graziers may no longer sound as angry as they once did, '...but when he's hand-feeding stock in a five-year drought and his gaze lifts to the high country—you'll still see it in his eyes.'

Differences of opinion over environmental management also caused clashes between the National Parks and Wildlife Service and the Snowy Mountains Authority.

Strict environmental guidelines built into the Snowy Mountains Hydro-Electric Power Act obligated the Authority to take whatever steps necessary to prevent erosion, siltation, bushfires and needless destruction of timber. The Authority worked hard from the beginning to minimise its environmental impact and to restore construction sites as quickly as possible. By 1955 its budget for soil conservation alone was 200 000 pounds. It pioneered techniques for reseeding plant cover over steep bedrock slopes. However, some of its measures were never going to be accepted by a parks administration which had set out to make the Kosciusko National Park a pristine wilderness.

In 1965, when the Island Bend–Guthega project was finished, the town was pulled down, ploughed up and seeded with native grasses. The Authority also planted willow trees and poplars to hold embankments in place to prevent erosion. But shortly after the completion of this work the superintendent of the park, Neville Gare, protested that the trees were not native. He wrote to Sir William Hudson asking him to have them pulled out because they were not in keeping with the natural format planned for the area.

Hudson wrote back brusquely, telling him not to be silly, and pointing out they were the only trees that would hold the banks in place. He suggested their presence was far better than having hillsides collapse with the first rains. Gare telephoned Hudson and repeated his demand. While Hudson was taking the call, one of the Authority's engineers, Tom Lewis, was in the office. By curious coincidence, the New South Wales minister responsible for the park was also named Tom Lewis.

As the parks superintendent delivered his ultimatum, Hudson casually leaned back in his chair and said, 'Well, I don't know if your ultimate boss would agree with you. As a matter of fact I've got Tom Lewis in the office with me now.' Turning to his engineer and speaking loudly enough for Gare to hear, he asked, 'What do you think about that stupid idea of removing all those trees from the national park, Tom?' The Authority's Tom Lewis responded, quietly, that indeed it was silly.

'There you are, he says it's a stupid idea too; so forget all about it,' Sir William told the superintendent.

Naturally, it did not take Gare long to discover he had been duped, and the flow of letters began again, with the National Parks eventually sending an ultimatum—they would pull out the trees and bill the Authority for the cost. Hudson discussed the problem with his senior engineer, Ken Andrews, who suggested the commissioner write back saying the Authority would uproot the trees as requested—the day after the National Parks removed the poplars at the ranger's house at Yarrangobilly Caves.

Hudson allowed himself a rare grin, sent the letter—and no further word was heard about the matter.

CHAPTER THIRTEEN

Ceremony marking the opening of Tumut–3 power station, 24 October 1972, an event which signalled the imminent completion of the Snowy Mountains Hydro-Electric Scheme.
PHOTO: NATIONAL LIBRARY CANBERRA

AFTERMATH

AFTERMATH

In the months immediately following the federal government's enactment of the Snowy Mountains Hydro-Electric Power Act in 1949 the Snowy Mountains Authority undertook a series of intensive studies in which it refined the original concept. The original sixteen power stations were reduced to seven but the generating capacity was lifted from 2620 to 3740 megawatts. The planned 800 kilometres of open-channel aqueducts were reduced to eighty kilometres of buried pipelines; yet by rearranging and linking storages the Authority lifted storage capacity by sixty per cent.

The primary function of the power stations is to meet peak demand for electricity in New South Wales, Victoria and the Australian Capital Territory, which is supplied through the New South Wales system. Broadly, peak demand is the period from 6 a.m. to 10 a.m. and from 5 p.m. to 9 p.m. The generating period for the hydro stations is flexible over a short term but, because of the channel capacity of rivers below the hydro stations and irrigation needs, is unable to replace base load generation by the thermal stations for prolonged periods.

The hydro stations operate in conjunction with state thermal stations, which burn black coal in New South Wales and mainly brown coal in Victoria. Because the transmission lines from the Snowy link the two states, energy is interchanged in what becomes effectively a single integrated system. This has considerable economic advantages. It allows power reserves to be shared and, in turn, optimises system costs.

When the Snowy Scheme reached its designed generating capacity of 3740 megawatts in 1974, this represented thirty-three per cent of the combined generating capacity of New South Wales, Victoria and the Snowy systems. Today the Snowy generates about sixteen per cent of total electricity produced.

One of the main advantages of hydro-electricity is its rapid response to changes in load. It is, for example, largely thanks to the Snowy that, in an industrial age powered by electricity, Australians can still instantly 'down tools' at two o'clock on the first Tuesday of each November to listen to or view the Melbourne Cup. Thermal power stations, which take some time to adjust output, would not be able to respond to the sudden drop in consumption. Hydro-electric power stations, however, can respond almost instantly to changes in load. It is simply a matter of turning on a big valve and releasing water to the turbines, or of shutting them down.

On Melbourne Cup day the thermal power stations in eastern Australia hold down their output, using the Snowy scheme to keep the system at maximum load. Shortly before race time, when the consumption begins to drop, the hydro stations simultaneously ease back. Similarly, when workplace equipment is turned on again after the race the hydro stations pick up the load, maintaining generation until the thermal stations can take over again.

Water for the Snowy's seven power stations is impounded by sixteen dams fed by their own catchments and from other catchments via the aqueducts. Like arteries in a huge heart, the 145 kilometres of tunnels shift the water about the Scheme to maintain a balance of supply and storage.

After passing through the power stations the waters, diverted inland from the Snowy and Eucumbene rivers to the Murray and Murrumbidgee catchments, are available for irrigation. In the final tailwater bays the water has fallen from an elevation of more than

1100 metres to just under 300 metres.

Through diversion, regulation and control of the rivers, the Scheme makes available for irrigation in the Murray and Murrumbidgee valleys an additional 2 360 000 megalitres of water over and above what might be harnessed naturally.

Whether one agrees with irrigation or not, the regulation of water provided by the Scheme played an important part in lessening the effect of severe droughts in 1967–68 and in 1982–83 in the irrigation areas. In 1982–83 the storage in the Scheme was reduced to eighteen per cent of capacity as it poured water into the parched farming areas. By mid-1989 the Scheme storage was back to more than sixty per cent of capacity.

During the two decades of construction, the numerous projects, each vast in its own right, overlapped and created an often confusing picture as to what precisely was being put together. For many it was like a gigantic jigsaw: lots of pieces, but no picture. For this reason the brilliance of the engineers who put it all together was often overlooked.

Because of the high profile of the first contractors and advisors from the United States there is also a misconception that Americans built the Scheme. In reality, more work was done by Australian than by any other group of contractors. All the pipelines and aqueducts were built by Australians; Monier built the Jindabyne–Island Bend tunnel as well as a number of other major projects; Thiess Brothers constructed the largest dam of the Scheme, Talbingo, which is also the biggest dam in the southern hemisphere.

The Scheme and its functions can be more clearly understood if each of the two main developments is viewed separately.

The Snowy–Tumut development is based on four power stations in the Tumut valley. The stations, two of which are deep underground, have a total generating capacity of 2180 megawatts, activated by the controlled waters of the Tumut, Tooma, Eucumbene and upper Murrumbidgee rivers. Water which is surplus to the generating requirements is conserved in Lake Eucumbene and any deficiency is drawn from the storage through the Eucumbene–Tumut tunnel.

The water required for electricity generation is routed first to the Tumut Pond Dam, where a slim concrete arch dam holds up the water for the Tumut-1 underground power station. Water is conveyed from Tumut Pond Dam to the power station by tunnel and shafts and, after passing through the turbines, is taken back to the Tumut River via a tailwater tunnel to Tumut-2 pondage.

The Tumut-2 power station, located 244 metres underground, is similar to Tumut-1, except that the tunnels are much longer. The tailwater tunnel from Tumut-2 discharges to the upstream end of Talbingo Dam, which is the storage for the 1500-megawatt Tumut-3 power station—the largest of the Scheme. Tumut-3 is a surface station, sited just below Talbingo Dam, which holds up the water to create the head for the station's turbines.

Tumut-3 power station operates on a pumped-storage principle. The water released after passing through its turbines is collected in the Jounama Dam for pumping back up into Talbingo Dam. This way the water can be recycled. Ultimately, however, it is fed further downstream into Blowering Dam, which feeds the Blowering power station near the town of Tumut. Blowering Dam holds the water released by all the upstream power stations until it is needed in the Murrumbidgee Irrigation Area.

This is necessary because the demand patterns for electricity and irrigation are quite

AFTERMATH

A worker surveys his handiwork, Tumut Pond.

Murray-1 power station.

different. Releases for electricity are higher in winter, when there is little or no irrigation. Conversely, irrigation needs are highest during February and March, when the electricity demand is relatively low.

The Snowy–Murray Development has a generating capacity of 1560 megawatts in three stations—Guthega, Murray-1 and Murray-2.

The Guthega power station is isolated from the rest. Situated at an elevation of 1325 metres on the upper Snowy River, it is the highest on the Scheme. Finished in 1955, it was the first power station to be completed. Its completion marked a turning point in the political fortunes of the Scheme.

Guthega is supplied from Guthega Dam through a 4.6-kilometre tunnel and two large pressure pipes. Water discharged from Guthega passes down the Snowy River for a short distance to Island Bend Dam. From there it can either be diverted inland through the Snowy–Geehi tunnel; or it can be conveyed through the Eucumbene–Snowy tunnel for storage in Lake Eucumbene, and returned when needed.

The Snowy–Geehi tunnel also accepts water pumped from Lake Jindabyne, and delivers all diverted flows to Geehi Dam. Geehi Dam also collects the flow of the Geehi River and is the pondage for Murray-1 power station. The water from Geehi flows to Murray-1 through an 11.8-kilometre pressure tunnel and three large steel pipelines. The discharge from Murray-1 continues on to the Murray-2 pondage which is formed by a concrete arch dam on Khancoban Back Creek.

Murray-2 power station in turn discharges into the Khancoban pondage, formed by a low dam across the Swampy Plain River. This pondage regulates the flow into the

AFTERMATH

Murray River, which in turn is regulated for Murray valley irrigation by the Hume Dam above Albury.

By 1967 the construction of the Scheme had passed the halfway mark. Three power stations—Guthega, Tumut-1 and Tumut-2—were operating and a fourth, Murray-1, was about to be commissioned. All the tunnels except Jindabyne–Island Bend and Murray-2 pressure tunnel were finished, as were most of the dams. All that remained were the Murray-2 project, the final stages of the Jindabyne works, the Tumut-3 project and Blowering.

In 1967 the American Society of Engineers rated the Snowy Scheme as one of the seven engineering wonders of the modern world. Sir William Hudson, the man whose tireless enthusiasm and unremitting demands had achieved this remarkable effort, had every reason to feel proud.

Eucumbene Dam on completion.

He was now seventy-two years of age and his tenure as commissioner had been extended twice by a special Act of Parliament. He remained as robust and keen as the day he took up the appointment in August 1949. He rarely relaxed, apart from Sunday afternoon strolls through the hills behind his house in Cooma. Even then he could be seen scribbling enthusiastically in a notebook plucked repeatedly from a breast pocket.

But in 1966 Sir Robert Menzies retired as Prime Minister and, in doing so, abandoned Hudson to a political wolfpack. Menzies was his strongest, and last, political ally. Despite assurances that he would be allowed to remain with the Scheme for as long as he was capable, Menzies' departure cleared the way for Hudson's political enemies to cut him down.

Many leading political figures had resented his close relationship with and influence on Menzies. One man in particular with whom Hudson had never seen eye to eye was David Fairbairn, who replaced Senator Sir William Spooner as Minister for National Development. When Spooner died in 1966, Hudson lost not only the other half of a long-running two-man team, but also a close friend. They had both served in France during World War I and Hudson was a pallbearer at Spooner's funeral. To be fair to Fairbairn, it would probably have been impossible for any replacement to gain Hudson's approval.

Walter Hartwig recognised the danger signs for Hudson when, at the breakthrough ceremony of one of the Island Bend tunnels, Hudson kept referring to Fairbairn as 'our new Minister', making it clear he considered him a Johnny-come-lately. 'I could see then there was trouble on the way,' said Hartwig.

With Menzies safely retired, Fairbairn saw to it that Hudson's term was not extended a third time. Hudson had been pressing vigorously for the Authority to be deployed on harnessing the large northern Queensland rivers for hydro-electricity. With adequate power, Hudson believed Australia could achieve anything. Spooner had shared this vision and was the main advocate in Canberra for the development of the Ord River Scheme. Fairbairn, on the other hand, talked of dismantling the Authority at the completion of the Snowy project.

The more Hudson lobbied to utilise the Authority's expertise in developing other water resources, the more Fairbairn talked of dismantling the organisation, thus increasing Hudson's frustration and sense of betrayal. Fairbairn was keen on a proposal put forward by two big Australian companies, Broken Hill Proprietary Limited (BHP) and Conzinc Rio Tinto Australia (CRA), to set up a private international engineering consultancy using Snowy staff.

Ultimately the government decided to retain the Authority to maintain the Scheme and set up its own consultancy, the Snowy Mountains Engineering Corporation; but this was largely as a result of pressure from the Department of External Affairs which had a number of Snowy engineers deployed on aid projects in developing countries under the Colombo Plan. It could see an important future for aid programs, particularly in South-East Asia, and lobbied hard for the retention of the Authority's expertise. (The Snowy Mountains Engineering Corporation is today imparting Snowy expertise in twenty-eight developing countries.)

By that time, however, Hudson was gone. He was retired in April 1967 on the eve

Sir William Hudson in the Murrumbidgee–Eucumbene tunnel, 1960.

of his seventy-second birthday. Despite two functions in his honour on his final day, he still put in a twelve-hour working day, finally leaving his office at midnight—just as he had done for most of the previous eighteen years.

His last function was a picnic on the banks of Lake Eucumbene with other Snowy staff who had joined in 1949. He told the dozens of reporters who trekked to Cooma for the event that the day was like all others, the only difference being that he would not be there tomorrow. He admitted to feeling 'a bit sad', and told journalists that he feared for the future of the Authority. He reminded them that it had on board some of the best scientific and engineering brains in the country and made a final, public plea for the Authority not to be dismantled.

Sir William listed as his two proudest achievements the slowing of the accident rate to half what it had been ten years earlier and the dramatic decline in road fatalities on the Scheme after the Authority's pioneering of seatbelts.

He also offered a few words of wisdom based on the Authority's experience with industrial relations:

> In Australia generally, they [industrial relations] are not good because there is a tremendous amount of suspicion and mistrust between management and the men. The workers are not always blameless, but we should not forget that management is often to blame as well. Employers have become too complacent. They must remember that the worker has to be considered when decisions are taken.
>
> We had a three-week strike at the beginning of our operation and I can tell you now that it was seventy-five per cent the fault of management.

His final comment was, 'I shall miss the mountains.'

Sir William moved to Canberra to work as a consultant for a British engineering firm but those who were close to him could see that his heart had been broken by the forced retirement so close to completion of the Scheme.

So shabbily was he treated by the new political masters that he did not even receive an invitation to the official opening ceremony of Murray-2 power station, which was commissioned just a few weeks after his retirement. The sting of this final political dart hurt him deeply and he slipped quietly into obscurity, a broken man.

Sir William Hudson died in 1978 at the age of eighty-two. The most apt final words on this remarkable man are those in a *Sydney Morning Herald* editorial of 28 April 1967, written to mark his retirement: 'If it can be said of any man that he has changed the ancient face of Australia, it can most assuredly be said of Sir William Hudson, engineer extraordinaire, administrator and father figure to thousands of workers from many, many countries.'

As the mighty construction program drew to an ordered finale under the administration of Hudson's former deputy, Howard Dan, time also caught up with the many other characters and personalities who gave birth to the Scheme and who, with either their brains or their hands, shaped its final form.

Major Hugh Powell Gough Clews finally retired in 1957 at the age of sixty-seven. His initial five-year contract had been extended to seven and a half years. He had given up his dream of a Blue Mountains hideaway and decided instead to retire in seclusion at Indi, the site he had first chosen as the main survey camp for the Geehi–Murray area.

He built a simple two-room mud-brick cottage. He had asked and was granted a lease, from Sir William Hudson, but went back to the commissioner when he realised the lease was only for thirty years. 'What do I do then?' he asked. Sir William took his old friend's shoulder and said, 'Come back and I'll renew it for another thirty.'

At his Indi sanctuary, the major indulged his passion for flowers and built a glasshouse to breed geraniums and miniature dahlias. He was also a prolific reader and poet and took a keen interest in the debate that was raging over whether Paul Strzelecki had been on the correct peak when he named Mount Kosciusko.

Some years after Strzelecki's ascent to the summit of the Australian Alps, his identification of Mount Kosciusko was challenged. The controversy persists to this day. Some believe his bearings were wrong and that the peak he referred to as the highest was Mount Townsend—the one he thought he was standing on—which is nineteen metres lower than the 2229-metre Mount Kosciusko.

The challenge was made by later explorers who climbed by a similar route and who argued that Strzelecki's description did not fit. Major Clews believed Strzelecki was correct and decided to use his experience as a surveyor in the area to prove it. He wrote a detailed and authoritative paper supporting Strzelecki, which did much to quell the controversy.

He often sat late into the night reading and writing by the light of a kerosene lamp. He asked Hildegard Martynow, wife of one of his first migrant chainmen, Kon Martynow, to make him black curtains so his light wouldn't disturb the nocturnal bush creatures.

The bush animals, and in particular a large family of kangaroos, became Major Clews's extended family. He fed them bread and made them feel safe and welcome. It was, after all, their pad he was on. For some time after he had left, people visiting the cottage would still be confronted by old kangaroos looking for handouts.

Happy at having finally found his solitude, Major Clews rarely left his camp. He kept in contact with friends in the outside world by radio and had groceries and supplies delivered by a local farmer, Mr Roger Mouat. Each dusk he greeted the end of another day with a tot of Lownes rum.

In 1978, at the insistence of friends who were worried about his advancing age and deteriorating health, he left his little cottage in the beautiful Indi Valley and moved into the Khancoban caravan park, to which he took an instant dislike. He fell ill about a year later and in 1980 was taken to hospital in Melbourne. He died there just before reaching his ninetieth birthday.

The Major's chainman, Kon Martynow, is today retired and living in Canberra. So too are many former Snowy workers: Jock Wilson; Ulick O'Boyle; the ballet dancer turned butcher, Jon Cambell; Dieter Amelung; and Bev Wales, who still leads an active life as a private investigator.

Few people are aware that, for most of his sixteen years in the mountains, Bev Wales was in constant pain from a back injury. Colleagues sometimes found him stretched on the floor in agony, yet never once did he complain or shirk what he saw as his duty. His only comment in later years was, 'It was that rotten Landrover that did it, jolting over those wretched tracks for sixteen years.'

Bev Wales, 1988.

Cooma, 1989. In the lower foreground is the railway station, the final stepping-off point for journeys from across the world.

The former packhorse handler, Max Paterick, today lives in Gosford; the German engineer, Walter Hartwig, lives in retirement in Sydney; so do Ivan Kobal and former Cooma detective, Bill Holmes. Boyd Mould fulfilled his ambition to remain a grazier on the Monaro, but without regaining access to the high country plains.

Tunneller Alessandro Wialletton runs his own garage in Cooma. Alessandro, or Sandro as he is now known locally, looks around the town today and sees that a good proportion of the more successful businessmen are former migrant workers who arrived on the Snowy with nothing: 'A lot of parents forbade their daughters having any contact with us. But look around Cooma today; the Australians are the workers and the migrants own and run the businesses. Today a lot of those girls' parents would have wished they'd let their daughters marry migrants.'

Jack Lawson, the young engineer who sat waiting to be burned alive by striking Yugoslavs, is today Professor of Civil Engineering at Melbourne University. Bob Leech, the wide-eyed youngster at Island Bend and later the determined parks ranger, is now with the Snowy River Shire Council. Others, such as Frank Rodwell, the tunnels inspector Colin Purcell, Neville Phee and Len Neithe, still work for the Authority. Those who have come down from the mountains to retire in Cooma include the former major contracts engineer, Ken Andrews; Ksenia Nasielski; Alf and Guiditta Fabbro and Mario and Angelina Pighin.

In a quiet street not far from where he worked at the Polo Flat workshops, the former Stuka pilot, Hein Bergerhausen, enjoys his retirement as he works at perfecting the arts of calligraphy and home winemaking. Like all the migrant workers who stayed, Hein is a proud Australian. He sought naturalisation the moment his five years of residency had passed.

Along with all the Europeans on the Snowy, Hein and Sybille Bergerhausen worked hard to become Australian citizens—'much harder than do native-born Australians, who are lucky enough to be born here.' But Hein also points out a difference between the early postwar migrants and others who followed later: 'We tried to settle quickly. In Cooma we didn't form a German Club like others later did in the cities. In Cooma we all became Australians.'

Hein left the Authority in 1956, disappointed at not securing a staff position that fell vacant in the mechanical engineering section. He joined the Allis Chalmers company, which had a maintenance contract with the Kaiser consortium.

In 1957 Kaiser decided to do its own servicing and appointed Hein as its mechanical supervisor. By August 1958 he had become the company's senior mechanical superintendent, responsible for the maintenance of all earthmoving equipment and controlling a staff of more than a hundred men.

In 1959, at the completion of the Eucumbene–Tumut project, Hein left Kaiser and bought the lease to the Cooma Caltex service station, which he operated for the next fifteen years—as well as a tyre recapping business and the agency for Peugeot, Renault and Mercedes-Benz cars.

In 1969 he joined the local Rotary Club, where he was also talked into joining the Ex-Servicemen's Club: 'How can I? I'm a square-head, a German,' he protested. But he was outnumbered by four Australians. 'It doesn't matter,' they said, and signed him in. Hein was glad they did: 'I got teased a bit because they were the winners but we all had some good times there. The Cooma RSL had a lot of new Australians, from both sides of the war, among its members. We were all ex-soldiers, that's all that mattered.'

But one thing Hein could not be talked into doing was marching on Anzac Day: 'I felt it was better to forget that side of it. Also I believed Anzac Day was a special day for Australians. It should be theirs alone to commemorate; but I'd join them after at the two-up.'

From 1970 to 1977 he initiated and organised beerfests at the local showgrounds. Featuring fifty-eight beers from around the world, the event drew crowds from throughout south-eastern Australia. 'The money raised—an average of 5000 dollars each year—gave me a chance to put something back into Cooma,' he said.

Sybille and Hein Bergerhausen in Cooma, 1989: 'We came looking for peace and found it.'

In 1970 Hein returned to Germany to visit his family and meet up with the men he had flown with during the war. It was a memorable reunion which was not without a touch of the surreal. Hein was now an Australian, and respectfully acknowledged as such by his former comrades-in-arms.

Nelson Lemmon, the former Minister for Works, died on 20 March 1989, at the age of eighty. He was the last link with the Chifley government which courageously, but to its own political cost, pushed an uncertain Australia into the industrial world.

In a newspaper interview (*Sydney Morning Herald*, 7 September 1987) Nelson Lemmon made a few biting comments about some of his political contemporaries. But he had only praise for Ben Chifley, whom he described as 'a leader of leaders, a man among men. He had charisma, if that's what you call it. No real factions when he was leader because everyone owed allegiance to him. You loved him and you worked for him.'

Today little remains of the settlements where men who built not just dams, but the foundations of a new nation, lived, worked, dreamed and died.

In 1989 the Snowy Mountains Authority was trying to obtain approval to mark the sites of the former construction camps and towns as places of historical significance but was facing determined opposition from the New South Wales National Parks and Wildlife Service.

Today there is strong disagreement about what is best for the mountains. Graziers still argue that sheep and cattle should be utilised as management tools to control the density of undergrowth and reduce the risk of fire. Stock leases were burned each year by a fire control officer after the snow had melted. Because the snow pushed the dead grass flat the fire sneaked over the ground and cleared the way for new shoots to emerge. Now that the grass is no longer burnt, graziers say it has formed an impenetrable mat over the ground and made it more difficult for new shoots to break through. The consequence, they say, is reduced ground cover and increased erosion.

It is one example of the differences of opinion that separate people with interests in the mountains. According to the National Parks and Wildlife Service the practice of burning was a prime cause of erosion. They point out that the dead grass protected the new shoots, and burning it removed that protection.

Graziers are also becoming increasingly angry over the expanding populations of feral animals in the national park, particularly wild dogs which, apart from savaging flocks on adjacent farmland, adversely affect native fauna. Under the lease agreements graziers were made responsible for controlling pests such as rabbits and foxes, which competed against both domestic and native animals, for keeping fences in good repair and maintaining the huts which had to be kept unlocked and stocked with emergency provisions.

The National Parks and Wildlife Service blames stockmen and the former inhabitants of Snowy camps and towns for the introduction of wild dogs and noxious weeds, and says it is doing its best to rid the park of the problems.

These conflicts have led to alienation and resentment in the mountains. Some would argue that they are symptomatic of the way Australians have forgotten many of the lessons learned during the construction of the Snowy and that they are pointers to a wider malaise in Australian society. The Snowy Scheme gave the nation the potential to achieve greatness; but the opportunity has been frittered away by lacklustre leadership and decades of government stagnation.

The Snowy Scheme generated enormous momentum; but Hudson's urgings that this should be harnessed for the development of northern Australia and for the establishment of an industrial base, using the world-class expertise available in the Authority, fell on deaf ears. Rather than build on the achievement and ethic of the Snowy, politicians reacted with childish spite against its success—and against the man who made it a success.

Almost from the moment the Scheme was finished, some politicians did their best to remove it from the national consciousness. Like bogong moths they went into a state of diapause—letting the country go back to living off its fatty reserves of wool, wheat and minerals while its new-found industrial, administrative and financial infrastructures slid back into limbo. It is for this stunted vision that Australians are paying today. It is the underlying factor behind the threat of a 'banana republic' and the crushing burden of high interest rates that result from an ever worsening trade balance.

The Snowy Scheme has been described in other countries as a monument to political vision, social harmony and engineering excellence. In Australia today it is a faded memory. When it does emerge, it is more often than not as a target. It has been attacked as a costly political stunt; as aiding and abetting irrigation, which some economists argue is an intrinsically uneconomic form of agriculture; and as being unnecessary because there is 'today' an over-capacity of electricity production, which in turn is inefficient and expensive.

In 1987–88, the average cost of energy production by the Scheme was 2.5 cents a Kilowatt-hour—half the cost of electricity sold to bulk consumers by the Electricity Commissions of New South Wales and Victoria.

In 1963, when the Scheme was still under construction, a leading economist, Professor A.J. Rose, estimated that while the Snowy was probably the cheapest means of providing peak load power when it was first designed, technological advances in coal-burning stations and in coalmining had reversed this position. The Authority responded that Professor Rose had failed to allow for the rapid depreciation of thermal plants, which have a life

The Authority's scientific services division pioneered a number of techniques for land rehabilitation. Here it is using a combination of straw matting and box-work to regenerate vegetation over bare rock.

of only thirty-five years. The working life of a hydro plant is seventy-five years, while the economic life of the capital base of hydro-electricity—dams and tunnels—may span centuries.

Nevertheless, it has been another argument deployed against the Scheme—an argument that ignores such benefits as non-pollution, the control of floods, the enriching social developments and the creation of a world-class engineering and technical base. The value of this last factor alone is incalculable. Many also forget the enhanced recreation facilities in the Snowy mountains that were made possible by the Scheme.

Although the Snowy Scheme showed that Australia could achieve something many Australians believed it could not, there remain people of political influence who stalk Canberra's corridors muttering that it should never have been built. Some see it as having thwarted Australia's nuclear ambitions; some still smart from old political wounds; some see it as an environmental defeat; and some blame it for increasing salinity in the irrigation areas. Some of the claims are ridiculous; others would take years to debate, and even then reach no final or worthwhile conclusion.

In terms of the environment and salinity, the Scheme, feeding out a reliable and regular water supply, offers the one measure of control in the exploitation of fragile river systems.

It may be timely to recall the perceptive warning penned almost 150 years ago by Paul Strzelecki, after his excursion through the Gippsland area. He predicted that the land faced an agricultural calamity if the plants, forest and undergrowth, which normally sheltered the soil and conserved moisture, did not stop disappearing 'under the innumerable flocks and axes which the settlers had introduced'.

Australians as a whole have yet to learn to live with this land; to take only that which it is capable of giving. The Snowy Mountains Scheme is an example of the possibility. The men who built the Scheme understood the mountains for what they were and the natural, renewable resource they had to offer. Nelson Lemmon worked tirelessly to impress upon all who entered the awesome realm of ridges and gorges the need to preserve all of the natural environment not directly necessary to the Scheme.

The fragility of the alpine environment was well understood. Stringent environmental obligations were built into the Act governing the Scheme; and this was at a time when the word environmentalist was not even in the dictionary and engineers and bulldozer drivers were accustomed to a free reign. The Scheme pioneered responsible environmental attitudes in engineering.

Through the exhaustive hydrology work undertaken before construction began, the architects of the Scheme knew just how much the mountains had to give. From an engineering perspective, the Scheme could have been much bigger. There were plans initially for seventeen power stations, eight of them on the Tumut River alone. It was possible, but there were reservations about the environmental impact and the strain that would be imposed on water reserves during a prolonged dry period—such as occurred during the early 1980s. There was no need for an Environmental Impact Study. It was sheer commonsense.

A question often asked is: Could the Snowy Scheme be built today? It is highly unlikely. Given the rate of depletion of Australia's forests and wildnerness areas in the past fifteen years, environmental concerns alone would kill the Scheme.

But even more significant is the fact that today the seeds of bold ideas are cast upon much harder, stonier ground. Unlike the men and women who devoted their lives to the development of the Snowy, Australians today live in a world paralysed by political pragmatism, journalistic negativism and the demand for immediate rewards.

Australians are governed now by a federal parliament from which all but the trappings of democratic management have been stripped. The days of open, honest debate have been destroyed by a cynical, shallow manipulation of parliamentary principles. Were the Snowy Scheme to be mooted today, the ability of members from both sides to ask important questions and learn about what was being planned would be found lacking. A country which once promised so much, and which demonstrated to the world that it could be as advanced and as visionary as any other, seems to slip further towards mediocrity with the passing of each parliamentary session.

It is inconceivable that a modern-day prime minister would go on air, as Chifley did in 1949, not only to explain a project like the Snowy to the people, but to urge them to be inquiring and to follow the debate. Chifley actually believed that government was responsible to the people.

The Snowy was given a budget and a timetable, and every Australian was encouraged to follow the project—even to visit and judge its merits for themselves. Busloads of visitors were a pain in the neck for engineers and construction teams, but the openness of the project allayed suspicion and ensured that the government and the Authority remained fully accountable.

Despite its current woes and seeming lack of purpose Australia, for the thousands who came to build the Snowy and the tens of thousands who have migrated since, is still a country of promise. To them it remains a place where humanity has been given a chance to start again, where people from all nations can build a new country based on shared prosperity and social justice. It is a country open to new ideas and lateral thinking; a country which cares for its environment and the welfare of future generations; a country which, because of its diverse ethnic mix, is tolerant and compassionate.

To them, the men killed on and inside the Snowy Mountains were not slain in a useless war, but fell in the building of a new, vibrant nation.

Perhaps the incident which most tellingly portrays the spirit of the Snowy occurred in Cabramurra on Anzac Day 1963.

The ex-diggers in the town decided the time had come to hold their own dawn service and march there, rather than go all the way to Cooma or Sydney. For the first time the Germans, Italians, Austrians and Hungarians, all of whom were represented in the small town, were openly uncomfortable. Realising this, the Australians discussed the matter among themselves and decided that all former soldiers, regardless of which side they had fought on, were welcome to attend the dawn service and the remembrance service after the march.

Later that morning, as the haunting notes of the Last Post drifted down the valley, men in the faded remnants of uniforms from a dozen different armies gathered in silent prayer to commemorate fallen comrades.

Afterwards, they all filed into the wet canteen and got blind drunk.

EPILOGUE

The new high-tech headquarters of the Snowy Mountains
Hydro-Electric Authority in Cooma.

FUTURE DIRECTIONS

FUTURE DIRECTIONS

The story of the Snowy Mountains Scheme is unlikely ever to be finished. The Snowy Mountains Hydro-Electric Authority and the Snowy Mountains Engineering Corporation are still vibrant organisations continually developing new technology. The workers who built the Scheme built a facility and enterprise which will continue to function for many decades, perhaps even centuries.

Invisible to the thousands of holidaymakers and skiers who flock to the Alps on the annual snow pilgrimage—yet pulsing through the ether—futuristic technology is already taking the Snowy Scheme to the forefront of technologies in the fields of electronics, computers, and climatology.

Computers, running programs developed by the Authority's own engineers, now control the vast volume of water that shifts through the trans-mountain tunnels between the lakes and dams which feed the Scheme. Balancing water inflow, storage, electricity generation and irrigation outflow through the dams and power stations is a giant juggling act that continues twenty-four hours a day.

Information shared by the computers is beamed across the snow-capped mountains by microwave towers, the only visible evidence of the new technology. The computers collect data at periodic intervals—for example, reservoir levels every five minutes, or spillway and diversion gate levels every ten seconds—and adjustments are made instantly. That data is fed into a database for access by other computer functions, such as those monitoring generator loading. These functions match, instantaneously, the electricity generated with the power demand from the Victorian and New South Wales electricity commissions. This real-time response maximises hydro-electricity's ability to increase output in seconds, compared with the much longer time required by coal- and oil-fired power stations. As the load on the system changes, the computers actually adjust valve openings to increase or decrease the volume of water being introduced to the turbines.

In addition to the operational control, the computer system developed by the Authority forecasts future generating requirements, calculating how to meet these while allowing for the down-time needed to maintain and refurbish machinery that has now been spinning for as long as thirty-four years.

About $400 million will be spent over the next twenty years in the refurbishing and modernisation program that is crucial as the Scheme enters the third phase of its development, which will take it into the twenty-first century. (The first phase was the construction period from 1949 to 1974, and the second the production period from the 1960s to today.)

Changing weather patterns have also forced the Authority to deploy resources into atmospheric research, drawing on the most advanced overseas technology in an attempt to address the very serious problem of diminishing water reserves.

Water reserves by the end of 1988 had dropped to forty-five per cent of capacity, although increased summer and autumn rainfall saw a significant rise in reservoir levels by mid-1989. The recent trend, however, has seen a steadily decreasing run-off from winter rains and snow over the past decade. Another year or two of the same trend and the Scheme could be confronted by an insufficient water supply to operate at maximum output. At this stage nobody knows whether the mountains are in the grip of a climatic cycle which will soon pass, or whether it is the beginning of a long-term, or even permanent, weather change caused by the greenhouse effect.

In 1988 the Authority sought the services of an expatriate Australian, Dr Joe Warburton, Professor of Physics at the University of Nevada, and regarded as one of the world's foremost weather researchers. He spent 1988 and 1989 in Australia to head a study into the possibilities of cloud-seeding the Australian Alps to increase snowfall. Seeding is a technique by which impurities such as silver iodide are introduced into clouds, turning the water to ice and causing precipitation.

The study, jointly funded by the Authority, and the Victorian and New South Wales electricity and water resource commissions, used sophisticated equipment from the Desert Research Institute of the University of Nevada to analyse the snow-bearing characteristics of cloudmasses which accumulate over the mountains in winter.

While cloud-seeding was regarded by many in Australia as a worthless exercise after it failed in the 1970s and early 1980s to break droughts over the grain belts, Dr Warburton claims cloud-seeding—or to use his preferred term 'weather modification'—is a proven technique. 'The only problem', he says, 'is one of statistics. Because of the natural variability of cloud you can't get ninety-nine per cent assurity.'

Dr Warburton confronted the critics, saying the Commonwealth Scientific and Industrial Research Organisation (CSIRO) abandoned cloud-seeding because the results were inconclusive as a drought cure for agriculture—not because the technique lacked scientific validity and did not work. (Most of Australia's earlier cloud-seeding efforts were over the dry interior, a region chosen on political rather than scientific grounds. The government of the day directed the effort according to influence exerted by the rural-based National Party. The program wasted millions of dollars. Experience in other countries demonstrates that cold cloudmass in areas where snow is the natural outcome is where seeding should be carried out.)

As well, the CSIRO, while it was a pioneer in basic cloud-seeding research, did not continue long enough to develop the type of technology and techniques that have since been developed in the United States where cloud-seeding is a common practice in many areas. There it is used to disperse cold fogs over airfields by making snow and clearing the atmosphere. It is also used in Canada to suppress hail in grain-growing areas.

Dr Warburton says cloud-seeding modifies, rather than changes, the weather. He scotches horror stories of accidental flood and tempest:

> Seeding increases precipitation by five to ten per cent. If the natural snowfall from a particular cloud mass would be twenty centimetres, the same cloud seeded would produce about twenty-two centimetres.
>
> It is an incremental increase over what would occur naturally. If the data suggests cloud-seeding would be viable it would simply augment natural precipitation. Also snow from seeded clouds tends to be powder snow, which is better for skiing.

The two-year research project in the ranges near Mount Kosciusko used the latest overseas advances in snow chemistry and remote sensing equipment. Two instruments, a short wavelength radar and a dual wavelength microwave radiometer, were used to profile the winter clouds that built up over the mountains.

Using chemistry techniques developed in the United States, the researchers also measured the isotopic composition of trace elements in the snow to obtain the background concentrations of naturally occurring impurities in the snowfall. This was to provide a measure of any subsequent changes caused by seeding. As a sideline it also checked the purity of the snow to see if it is being polluted as the weather passes over coastal industrial centres.

Another instrument, an ice crystal replicator, recorded the shape and size of snowflakes as they hit the ground. Analysis of the crystals determines the temperature at which they form and at what level in the cloud. The aim was to find the location in the cloud of liquid 'super-cooled' water. Before seeding can be considered scientists must learn the whereabouts of the water layer—it should preferably be low in the cloud—and whether there is enough of it to make seeding an economic proposition.

In Nevada, where cloud-seeding has become a fact of life, the clouds are seeded from the ground by devices installed in the mountains. When released, by remote control, silver iodide crystals, which are invisible to the naked eye, simply rise into the cloud. The Americans have found this to be more efficient and economic than seeding from aircraft.

A decision on whether to proceed with a full-scale cloud-seeding program in the Snowy is expected to be made some time in 1990. Ideally, the Authority would like not simply to recover lost water reserves, but actually to increase the amount of water at its disposal.

With a larger water base the Scheme could significantly increase its output by running the power stations beyond the morning and evening peak load periods. In the light of the country's ever-increasing energy demands, this will become an important political consideration as concern mounts over the pollution caused by coal-fired stations.

Appendix One
ROLL OF HONOUR

MEN KILLED WHILE AT WORK ON THE SCHEME

The following honour roll was compiled by the Authority in 1981 for a memorial in Cooma. Owing to factors such as the wide area over which accidents occurred, the jurisdiction of coroners courts in two states, and the incomplete records of many contractors, the list appears to be well short of the true death toll. A number of men known to have been killed in accidents are not listed here. An indication of the large gap in the records is given by the fact that the official toll is one hundred and twenty-one, yet Constable Bev Wales handled ninety-one inquests involving Snowy workers during his time at Khancoban alone. He estimates he investigated up to thirty more fatalities at his other postings; and he was just one of many policemen in many areas who were responsible for investigating construction fatalities. In addition to this, some men worked under pseudonyms, which at times made full identification difficult, resulting in gaps concerning nationality, religion and next of kin. The vehicle fatalities listed relate only to those accidents that occurred while the driver was on Authority business. Speed, alcohol, shocking roads and extreme weather conditions combined to kill dozens more in their time off. There was also a significant number of suicides ascribed to the lingering stresses of war trauma and loneliness. There was a number of attempted murders, but only three actual killings on the Scheme.

Date	Name	Nationality	Employer	Location	Listed Cause
5.3.52	Henriksen, N.	unstated*	Selmer Engineering	Guthega Dam	Crushed by falling rock
14.1.53	Henriksen, P.B.	unstated	Selmer Engineering	Guthega Dam	Crushed by falling rock
19.1.53	Pinazza, O.V.	Italy	SMHEA*	Jindabyne camp	Fall from power pole
13.4.53	Smolenski, M.	unstated	Selmer Engineering	Guthega tunnel	Struck by winch handle
8.9.53	Balzer, W.	unstated	Selmer Engineering	Guthega surge tank	Struck by falling rocks
1.10.53	Berg, J.	unstated	Selmer Engineering	Guthega tunnel	Struck by falling rocks
17.11.53	Degani, V.	Italy	SMHEA	Polo Flat, Cooma	Road accident
5.12.53	Tuite, T.B.	Britain	SMHEA	Tolbar Road	Tractor accident
19.1.54	Ulseth, R.	unstated	Selmer Engineering	Guthega tunnel	Struck by falling rocks
6.5.54	Brenycz, D.	Poland	SMHEA	Polo Flat, Cooma	Tractor accident
9.6.54	Wagner, B.	unstated	Selmer Engineering	Guthega power station	Fall from scaffolding
4.9.54	Ilic, M.	Yugoslavia	SMHEA	Polo Flat, Cooma	Shock from burns
14.5.55	Rullis, Peteris	unstated	Etudes et Entreprises	Happy Valley Road	Tractor accident
6.7.55	Farrell, R.M.	Australia	SMHEA	Tumut Pond	Vehicle accident
12.10.55	Symonds, R.M.	Australia	SMHEA	Coolringdon	Tractor accident
15.12.55	McAlley, Eric	unstated	KWPR*	Tumut Pond	Tractor accident
17.12.55	Cerubino, L.	unstated	KWPR*	Junction Shaft	Rock fall
7.4.56	Sellan, Angelo	Australia	KWPR	Junction Shaft	Rock fall
10.4.56	Dall'Antonia, Silvo	Italy	KWPR	Junction Shaft	Rock fall
10.4.56	Yannacopoulis, T.	Greece	KWPR	Junction Shaft	Rock fall
13.4.56	Buckman, H.J.	Australia	Etudes et Entreprises	Tumut River Road	Hit by falling tree
19.6.56	Reid, Alexander	Australia	KWPR	Eucumbene portal	Tunnel loco accident
14.8.56	Farrant, F.J.	Australia	SMHEA	Jindabyne camp	Road accident
7.12.56	Evans, Jack	Australia	KWPR	T1 pressure tunnel	Tunnel loco accident
15.12.56	Jonjoric, Eddie	Poland	KWPR	Junction Shaft	Rock fall

ROLL OF HONOUR

Date	Name	Nationality	Employer	Location	Cause
8.1.57	Herzberg, Valker	Germany	KWPR	Junction Shaft	Rock fall
4.3.57	Duane, Patrick J.	Australia	KWPR	Adaminaby Dam	Tractor accident
4.4.57	Perkovic, D.	Yugoslavia	SMHEA	Tumut Pond	Road accident
30.4.57	Miniutti, Marcello	NOK Aust*	KWPR	Adaminaby Dam	Tractor accident
18.5.57	Ilic, Vladimir	unknown	KWPR	Adaminaby Dam	Tractor accident
16.7.57	Savegnano, Luigi	Australia	KWPR	Junction Shaft	Tunnel loco accident
4.10.57	Beard, Nicholas J.	Britain	KWPR	Adaminaby Dam	Road accident
4.10.57	Mori, Bogemir	unknown	KWPR	Adaminaby Dam	Tractor accident
1.11.57	Koch, G.	Italy	SMHEA	Geehi	Drowned
7.1.58	Yong, N.	Holland	KWPR	Adaminaby Dam	Tractor accident
17.3.58	Pellegrino, D.	Italy	SMHEA	Nungar camp	Truck accident
18.3.58	Costello	unknown	KWPR	Adaminaby Dam	Tractor accident
16.4.58	Rugolo, G.	unknown	Etudes et Entreprises	Kenny's Knob	Fall down shaft
16.4.58	Fizzoll, B.	unknown	Etudes et Entreprises	Kenny's Knob	Fall down shaft
16.4.58	Di Salvio, M.	unknown	Etudes et Entreprises	Kenny's Knob	Fall down shaft
16.4.58	Gusela, M.A.	unknown	Etudes et Entreprises	Kenny's Knob	Fall down shaft
23.5.58	McDonald, John	Australia	KWPR	Junction Shaft	Heart attack
10.7.58	Brander, N.L.	Australia	SMHEA	Tooma Road	Truck accident
2.9.58	Adams, Garry	Australia	KWPR*	T2 Tunnel	Tractor accident
26.9.58	Van Hoof, Peter L.	Belgium	KPMR	Section Creek	Tractor accident
4.10.58	Walters, George	Australia	Hall & Elliot	Byatt's Hut	Tractor accident
9.10.58	Hallop, Clifford	Australia	Thiess Bros	Tooma Dam	Tractor accident
12.12.58	Cupido, Franchesir	Italy	Thiess Bros	Tooma Dam tunnel	Rockfall
9.1.59	De Vries, H.	Australia	Utah-Brown-Root	Tantangara	Crushing plant accident
2.3.59	Dramisino, S.	Italy	SMHEA	T1 surge chamber	Fall down shaft
9.3.59	Pagge, Alfonso	unknown	Thiess Bros	Deep Creek tunnel	Rockfall
29.4.59	Johansen, Ludwig	Norway	Utah-Brown-Root	Tantangara	Electrocuted
5.5.59	Dahlberg, Graham E.	Australia	Utah-Brown-Root	Providence heading	Tunnel loco accident
13.5.59	Fazey, Thomas	unknown	KPMR	T2 tailrace	Tunnel loco accident
30.7.59	Waldon, Frederick	Australia	Utah-Brown-Root	Providence heading	Premature explosion
30.7.59	Harodes, Leonid	Estonia	Utah-Brown-Root	Providence heading	Premature explosion
30.7.59	Darsie, Ivo	Italy	Utah-Brown-Root	Providence heading	Premature explosion
20.8.59	Maniatis, Peter	NOK Aust.	KPMR	River camp	Crushed by falling steel
9.10.59	Vettor, Giuseppe	NOK Aust.	KPMR	T2 tailrace	Tunnel loco accident
24.10.59	Vecchiato, Drenesto (Ernie)	Italy	KPMR	T2 draft tube	Explosion
2.11.59	Cacic, N.	Yugoslavia	Thiess Bros	Burn Creek shaft	Fall down shaft
5.11.59	Beccia, L.	Italy	Utah-Brown-Root	Tantangara Road	Tractor accident
17.11.59	Mangelle, Joseph	Austria	KPMR	Goat Ridge Road	Buried by fall of sand
17.3.60	Sledzianowski, H.	unknown	Theiss Bros	Deep Creek Road	Tractor accident
29.4.60	Fergusson, R.	Australia	SMHEA	Tolbar Road	Road accident
16.5.60	Hager, Joseph	Austria	Thiess Bros	Tooma gate shaft	Hit by falling timber
6.6.60	Biscontin, Tomaso	Italy	KPMR	T2 Dam site	Tractor accident
20.6.60	Nemeth, L.	Australia	Utah-Brown-Root	Providence heading	Hit by falling pipe
22.8.60	Gasparetto, Dahilo	NOK Aust.	Thiess Bros	Deep Creek Adit	Tunnel loco accident
17.1.61	Hirkic, Sharban	Yugoslavia	KPMR	T2 power station	Fall from scaffolding
12.2.61	Seaman, N.	Australia	Thiess Bros	Tooma Haul Road	Road accident
30.8.61	Dawson, A.	Australia	SMHEA	Leatherbarrel Creek	Crushed by falling load
2.4.62	Thayne, D.	NOK Aust.	Utah-Brown-Root	Nimmo sand deposit	Tractor accident
28.5.62	Thiess, Kenneth A.	Australia	Thiess Bros	Geehi Dam Road	Struck by dislodged log
1.8.62	Luisotto, M.	Australia	Utah-Brown-Root	Snowy–Euc. tunnel	Rockfall
27.8.62	Lascukowski, H.W.	unknown	Utah-Brown-Root	Snowy–Euc. tunnel	Tunnel loco accident
6.11.62	Longo, C.	Australia	Thiess Bros	Bogong adit	Tunnel loco accident
16.2.63	Winnicombe, J.	unknown	Utah-Brown-Root	Geehi control shaft	Rockfall
19.4.63	Argyroupoulos, C.	Greece	Utah-Brown-Root	Burrungubugge heading	Tunnel loco accident
28.4.63	Ackerman, P.	unknown	Utah-Brown-Root	Island Bend Dam	Fell from dam

Date	Name	Nationality	Employer	Location	Cause
6.8.63	Lafuente, A. (Pedro)	Spain	Thiess Bros	Snowy–Geehi tunnel	Rockfall
24.9.63	Kurent, E.	Yugoslavia	Thiess Bros	Snowy–Geehi tunnel	Struck by flying rock
18.10.63	Sramz, E.	Italy	Utah-Brown-Root	Burrungubugge heading	Rockfall
29.10.63	Miraliotis, Nicholas	Greece	Utah-Brown-Root	Geehi control shaft	Fell down shaft
16.12.63	Texic, Mirko	Yugoslavia	Thiess Bros	Geehi adit	Electrocuted
21.12.63	Mayorol, A.	Spain	Utah-Brown-Root	Geehi control shaft	Falling concrete pipe
21.12.63	Coranov, J.	Yugoslavia	Utah-Brown-Root	Geehi control shaft	Falling concrete pipe
21.12.63	Mezlyak, S.	Yugoslavia	Utah-Brown-Root	Geehi control shaft	Falling concrete pipe
31.1.64	Grigori, Peter	Rumania	Utah-Brown-Root	Burrungubugge heading	Rockfall
3.2.64	Brescacin, Elio	Italy	Utah-Brown-Root	Island Bend heading	Explosion
20.2.64	Scott, W.W.	Australia	Utah-Brown-Root	Island Bend heading	Rockfall
9.3.64	Breier, Christian	Germany	Thiess Bros	M1 pressure tunnel	Electrocuted
9.9.64	Oakley, Daniel	Ireland	Kaiser Engineers	Swampy Plains River	Drowned
15.9.64	Rieder, Walter	Austria	Thiess Bros	Geehi aqueduct tunnel	Tunnel loco accident
14.12.64	Egan, Michael	Ireland	Perini	M1 power station	Crane accident
4.1.65	Briska, J.	Britain	SMHEA	Jindabyne camp	Tractor accident
21.1.65	Nagy, J.B.	Hungary	Thiess Bros	Geehi Dam	Tractor accident
22.2.65	Yule, Gordon	Britain	Thiess Bros	Snowy–Geehi tunnel	Electrocuted
28.4.65	Perez, Augustin	Spain	Utah-Brown-Root	Burrungebugge heading	Tunnel loco accident
1.5.65	Kokles, Wayne A.	Australia	Thiess Bros	Geehi Dam	Tractor accident
14.6.65	Burns, David Allen	unknown	Kaiser Engineers	Khancoban	Disintegrated grindstone
15.8.65	Bullock, Neville J.	Australia	M.K.U.Mc*	Blowering Dam site	Tractor accident
28.12.65	Heyerma, Hendrick	Britain	Thiess Bros	Blowering Dam site	Road accident
21.12.65	Grodum, Olaf	Norway	Utah-Brown-Root	Cobber Creek quarry	Tractor accident
6.2.66	Kolari, S.	unknown	Thiess Bros	unknown	Crushed by falling steel
14.4.66	Hiscox, Brian J.	Australia	Brambles	Tooma–Khancoban road	Road accident
21.5.66	Diatchenko, Nickolas	Australia	Société Dumez	Talbingo Village	Electrocuted
10.6.66	Compton, E.	Australia	SMHEA	Geehi aqueduct	Truck accident
14.8.66	Bott, Serafino	Italy	Humes Ltd	unknown	Pipeline accident
26.9.66	Kouchouvelis, Peter	Greece	John Hollands Const.	unknown	Tractor accident
13.5.67	Satral, Louis	Hungary	Société Dumez	Jounama Dam	Tractor accident
25.8.67	Gaffy, Alan Thomas	Australia	Dillingham	unknown	Tractor accident
26.10.67	Lucas, Gordon C.	Australia	Doon Bros	unknown	Crushed by falling steel
9.8.68	Derrig, William	Ireland	Thiess Bros	Talbingo	Tractor accident
17.10.68	Covic, Peter	Croatia	Thiess Bros	unknown	Road accident
11.11.68	Svoboda, K.	Czechoslovakia	SMHEA	Near Adaminaby	Road accident
17.12.68	Kovacevic, L.	Albania	SMHEA	Near Adaminaby	Road accident
10.3.70	Barrow, Jeff	Australia	Thiess Bros	unknown	Killed by falling concrete
22.6.70	Watts, John	Australia	Thiess Bros	unknown	Killed by falling power pole
31.2.72	Kohler, H.	Germany	SMHEA	Guthega	Electrocuted

*Key:

Unstated	Nationality not stated, but most likely Norwegian
SMHEA	Snowy Mountains Hydro-Electric Authority
KWPR	Kaiser-Walsh-Perini-Raymond; an American consortium
NOK Aust.	Nationality unstated, but next of kin in Australia
KPMR	Kaiser-Perini-Morrison-Raymond
M.K.U.Mc	Morrison-Knudson of Australia, Utah Construction & Engineering, and McDonald Constructions.

Two Snowy workers were also killed while working for the Authority overseas. They were:

E.D. Compton, an Australian, killed in a tractor accident in Thailand; and

M. Lavrnja, from Yugoslavia, killed in a tractor accident at Bougainville, Papua New Guinea.

Appendix Two

TECHNICAL INFORMATION

SUMMARY OF PRINCIPAL FEATURES

Source: Snowy Mountains Hydro-Electric Authority, *Engineering Features of the Snowy Mountains Scheme.*

Tunnels

Tunnel	Length (km)	Excavated diameter (metres)	Year of completion
Eucumbene–Snowy	23.5	6.35	1965
Eucumbene–Tumut	22.2	6.91	1959
Murrumbidgee–Eucumbene	16.6	3.35	1961
Snowy–Geehi	14.5	6.30	1966
Tooma–Tumut	14.3	3.71	1961
Murray-1 pressure	11.8	7.44	1966
Tumut-2 pressure & tailwater	11.3	7.01	1961
Jindabyne–Island Bend	9.8	3.95	1968
Guthega	4.7	5.74	1955
Murray-2 pressure	2.4	8.08	1969
Tumut-1 pressure	2.4	7.01	1959
Tumut-1 tailwater	1.3	8.10	1959

Power Stations

Power Station	Installed capacity (kW)	Number of units	Rated head (metres)	In service
Tumut-3	1,500,000	6	150.9	1973
Murray-1	950,000	10	460.2	1967
Murray-2	550,000	4	264.3	1969
Tumut-1	320,000	4	292.6	1959
Tumut-2	280,000	4	262.1	1962
Blowering	80,000	1	86.6	1969
Guthega	60,000	2	246.9	1955

Pumping Stations

	Rated capacity (cumecs)	Pumping head (metres)	In service
Tumut-3	297.0	155.1	1973
Jindabyne	25.5	231.6	1969

Dams

Dam	Type*	Height (metres)	Crest length (metres)	Reservoir capacity (megalitres)	Year of completion
Talbingo	R	161.5	701.0	920,600	1970
Eucumbene	E	116.1	579.1	4,798,300	1958
Blowering	R	112.2	807.7	1,632,400	1968
Geehi	R	91.4	265.2	21,106	1966
Tumut Pond	CA	86.3	217.9	52,818	1959
Jindabyne	R	71.6	335.3	689,900	1967
Tooma	E	67.1	304.8	28,125	1961
Island Bend	CG	48.8	146.3	3,020	1965
Tumut-2	CG	46.3	118.9	2,700	1961
Tantangara	CG	45.1	216.4	254,100	1960
Jounama	R	43.9	518.2	43,800	1968
Murray-2	CA	42.7	131.1	2,300	1968
Guthega	CG	33.5	139.0	1,800	1955
Happy Jacks	CG	29.0	76.2	270	1959
Deep Creek	CG	21.3	54.9	11	1961
Khancoban	E	18.3	1066.8	21,512	1966

*R = Rockfill, E = Earthfill, CG = Concrete Gravity, CA = Concrete Arch

PRINCIPAL CONTRACTORS SNOWY–TUMUT DEVELOPMENT

Works constructed	Date of award of contract	Contractor	Value of contract at date awarded
Eucumbene Dam	May 1956	*Kaiser-Walsh-Perini-Raymond	
Eucumbene–Tumut tunnel & Happy Jacks Dam	April 1954		
Tumut Pond Dam & Tumut-1 pressure tunnel	April 1954		$38,000,000
Murrumbidgee–Eucumbene tunnel & Tantangara Dam	May 1958	Utah Australia Ltd & Brown and Root Sudamericana Ltd of USA	$12,800,000
Tooma Dam & Tooma–Tumut tunnel	May 1958	Thiess Brothers Ltd, Australia	$18,200,000
Tumut-1 power station, pressure shafts & tailwater tunnel	April 1954	**French Consortium under Etudes et Entreprises	$7,800,000
Tumut-1—generators	June 1954	ASEA Electric (Aust) Pty Ltd	$2,177,995
—turbines	May 1954	English Electric Co Ltd	$1,722,164
—transformers	August 1955	ACEC Charleroi, Belgium	$792,846
Tumut-2 Dam & Tumut-2 power station, pressure tunnel & tailwater tunnel	May 1958	***Kaiser-Perini-Morrison-Raymond	$35,800,000

TECHNICAL INFORMATION

Works constructed	Date of award of contract	Contractor	Value of contract at date awarded
Tumut 2—generators	December 1957	ASEA Electric (Aust) Pty Ltd	$2,339,400
—turbines	October 1957	Charmilles Eng. Works Ltd	$1,846,772
—transformers	April 1958	CA Parsons & Co Ltd	$902,104
Talbingo Dam & Tumut-3 power station	October 1967	Thiess Bros Pty Ltd Australia	$41,900,000
Tumut-3 pressure pipelines	February 1968	Humes Ltd, Australia / Sulzer Bros. Switzerland	$8,100,000
Tumut 3—generators	November 1966	Mitsubishi (Australia) Pty Ltd	$3,426,465
—turbines	November 1966	Tokyo Shibaura Electric Co Ltd	$3,655,261
—pumps	November 1966	Tokyo Shibaura Electric Co Ltd	$1,651,298
—transformers	December 1967	Tyree Electrical Co Pty Ltd	$1,465,267
Jounama Dam	March 1966	Société Dumez, France, operating as Dumez Australia Ltd	$4,100,000
Blowering Dam	March 1965	Morrison-Knudsen (Aust), Utah Construction & Engineering Pty Ltd & McDonald Constructions Pty Ltd	$21,600,000
Blowering power station & outlet works	September 1966	Morrison-Knudsen International, Utah Construction & Mining, & McDonald Constructions Pty Ltd	$3,100,000
Blowering—generator	June 1965	John Carruthers & Co Pty Ltd	$327,092
—turbine	June 1965	Riva-Calzoni	$523,708
—transformer	July 1966	Ercole Marelli & Co, Italy	$120,000

*Joint venture comprising Henry J. Kaiser Corp., Walsh Construction Co., Perini Corporation, Bates & Rogers Construction Corporation, The Arthur A. Johnson Corporation and RCP Construction Company Ltd—all of the USA.

**French group of contractors comprising Compagnie Industrielle de Travaux, Entreprise Fougerolle pour Travaux Publics, Société Générale d'Entreprises, Etudes et Entreprises, L'Entreprise Industrielle, and Société Nationale de Travaux Publics.

***Joint venture comprising Henry J. Kaiser Corporation, Perini Corporation, Morrison-Knudsen of Australia Ltd, RCP Construction Corporation Ltd and Bates and Rogers Construction Corporation—all of the USA.

PRINCIPAL CONTRACTORS
SNOWY–MURRAY DEVELOPMENT

Works constructed	Date of award of contract	Contractor	Value of contract at date awarded
Guthega Dam, Guthega tunnel, pressure pipeline, Guthega power station	September 1951	Ingenior F. Selmer A/S, Oslo Norway, operating as Selmer Engineering Pty Ltd	$12,000,000
Guthega—turbo-generators	April 1951	English Electric Co Ltd	$1,007,386
—transformers	March 1953	Hackbridge & Hewittic Ltd, England	$220,510

SNOWY

Works constructed	Date of award of contract	Contractor	Value of contract at date awarded
Eucumbene–Snowy tunnel, Snowy–Geehi tunnel (Snowy section) & Island Bend Dam	October 1961	Utah Australia Ltd & Brown and Root Sudamericana Ltd of the USA	$41,800,000
Jindabyne Dam	February 1965	Utah Australia Ltd & Brown and Root Sudamericana Ltd of the USA	$5,600,000
Jindabyne pumping station, pipeline & surge tank	January 1966	John Holland (Construction) Pty Ltd, Australia	$5,000,000
Jindabyne pumping station —pumps	July 1964	Escher Wyss Ltd	$904,406
—transformers	May 1965	AEI Engineering Pty Ltd	$144,054
—motors	July 1964	Siemens Halske Siemens Schuckert (Aust) Pty Ltd	$490,431
Jindabyne–Island Bend tunnel	May 1964	Monier-McNamara-Hardeman—a joint venture of Concrete Industries (Monier) Ltd of Australia, McNamara Corporation Ltd of Canada, and Paul Hardeman Inc of the USA	$8,200,000
Snowy–Geehi tunnel (Geehi section), Geehi Dam, Murray-1 pressure tunnel & surge tank	July 1961	Thiess Bros Pty Ltd Australia	$37,800,000
Murray-1 pressure pipelines	July 1962	Humes Ltd, Australia / Sulzer Bros, Switzerland	$4,400,000
Murray-1 power station	November 1962	Perini (Aust) Pty Ltd, an associate company of the Perini Corporation, Framingham, Mass., USA	$3,200,000
Murray-1—generators	August 1961	ASEA Electric (Aust) Pty ltd	$2,589,264
	September 1963		$666,014
—turbines	August 1961	Boving and Co (ANZ) Pty Ltd	$3,789,682
	June 1963		$969,844
—transformers	August 1961	Australian General Electric Co Pty Ltd	$839,710
	June 1964		$245,276
Murray-2 Dam (Stage 1)	December 1964	Thiess Bros Pty Ltd, Australia	$770,000
Murray-2 Dam (Stage 2), pressure tunnel & power station	December 1965	Dillingham Constructions Pty Ltd, Australia—an associate company of the Dillingham Corporation, Hawaii, USA	$14,900,000
Murray-2 pressure pipelines	May 1966	Humes Ltd, Australia / Sulzer Bros., Switzerland	$8,400,000
Murray-2—generators	January 1964	ASEA Electric (Aust) Pty Ltd	$1,599,476
—turbines	January 1964	Hitachi Limited	$2,052,058
—transformers	November 1964	Email Limited	$603,698
Khancoban Dam & Murray-2 power station excavation	August 1963	Kaiser Engineers and Construction, USA	$4,000,000

TECHNICAL INFORMATION

SNOWY MOUNTAINS SCHEME
CONSTRUCTION TIMETABLE

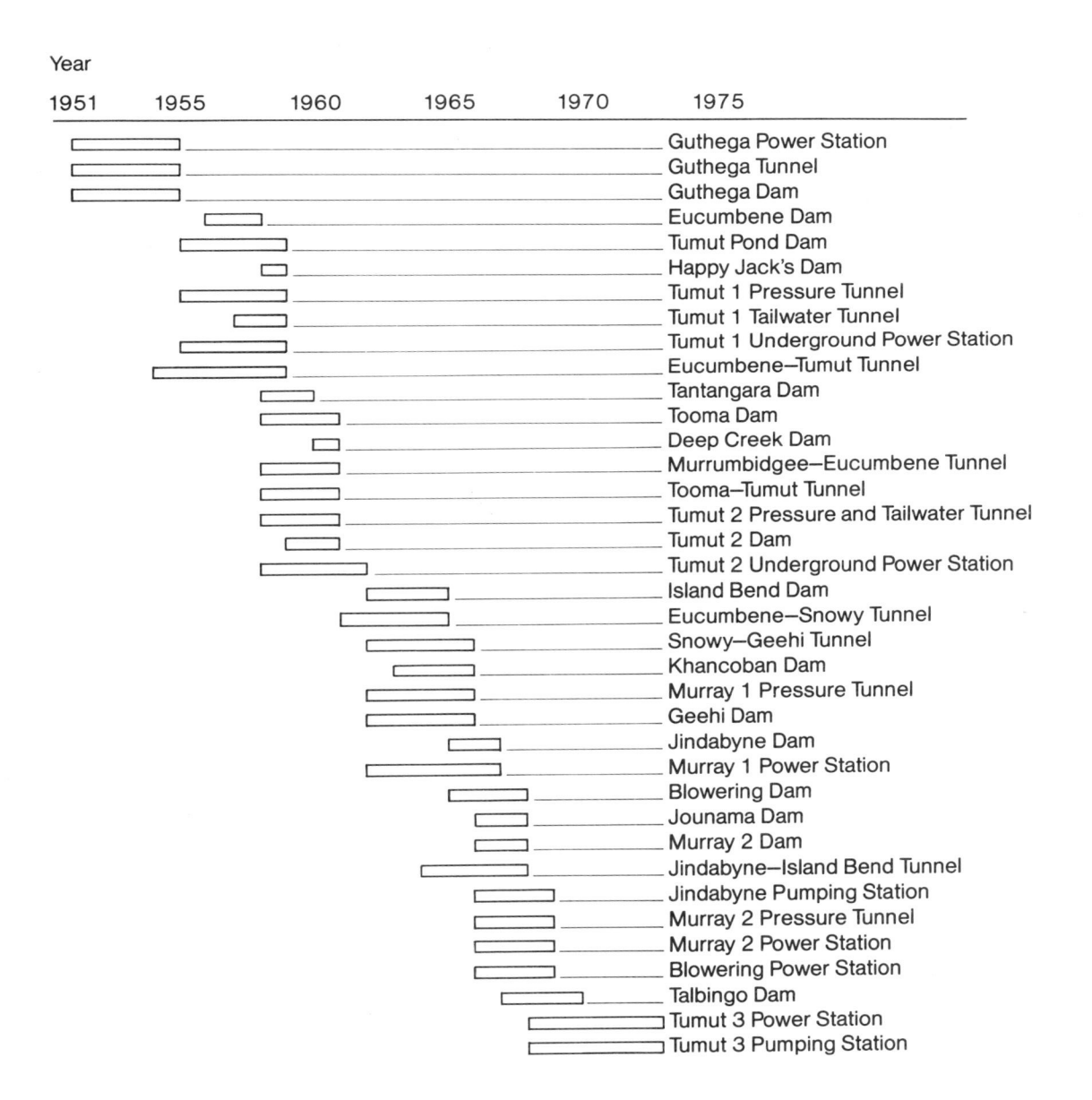

Year
1951   1955   1960   1965   1970   1975

- Guthega Power Station
- Guthega Tunnel
- Guthega Dam
- Eucumbene Dam
- Tumut Pond Dam
- Happy Jack's Dam
- Tumut 1 Pressure Tunnel
- Tumut 1 Tailwater Tunnel
- Tumut 1 Underground Power Station
- Eucumbene–Tumut Tunnel
- Tantangara Dam
- Tooma Dam
- Deep Creek Dam
- Murrumbidgee–Eucumbene Tunnel
- Tooma–Tumut Tunnel
- Tumut 2 Pressure and Tailwater Tunnel
- Tumut 2 Dam
- Tumut 2 Underground Power Station
- Island Bend Dam
- Eucumbene–Snowy Tunnel
- Snowy–Geehi Tunnel
- Khancoban Dam
- Murray 1 Pressure Tunnel
- Geehi Dam
- Jindabyne Dam
- Murray 1 Power Station
- Blowering Dam
- Jounama Dam
- Murray 2 Dam
- Jindabyne–Island Bend Tunnel
- Jindabyne Pumping Station
- Murray 2 Pressure Tunnel
- Murray 2 Power Station
- Blowering Power Station
- Talbingo Dam
- Tumut 3 Power Station
- Tumut 3 Pumping Station

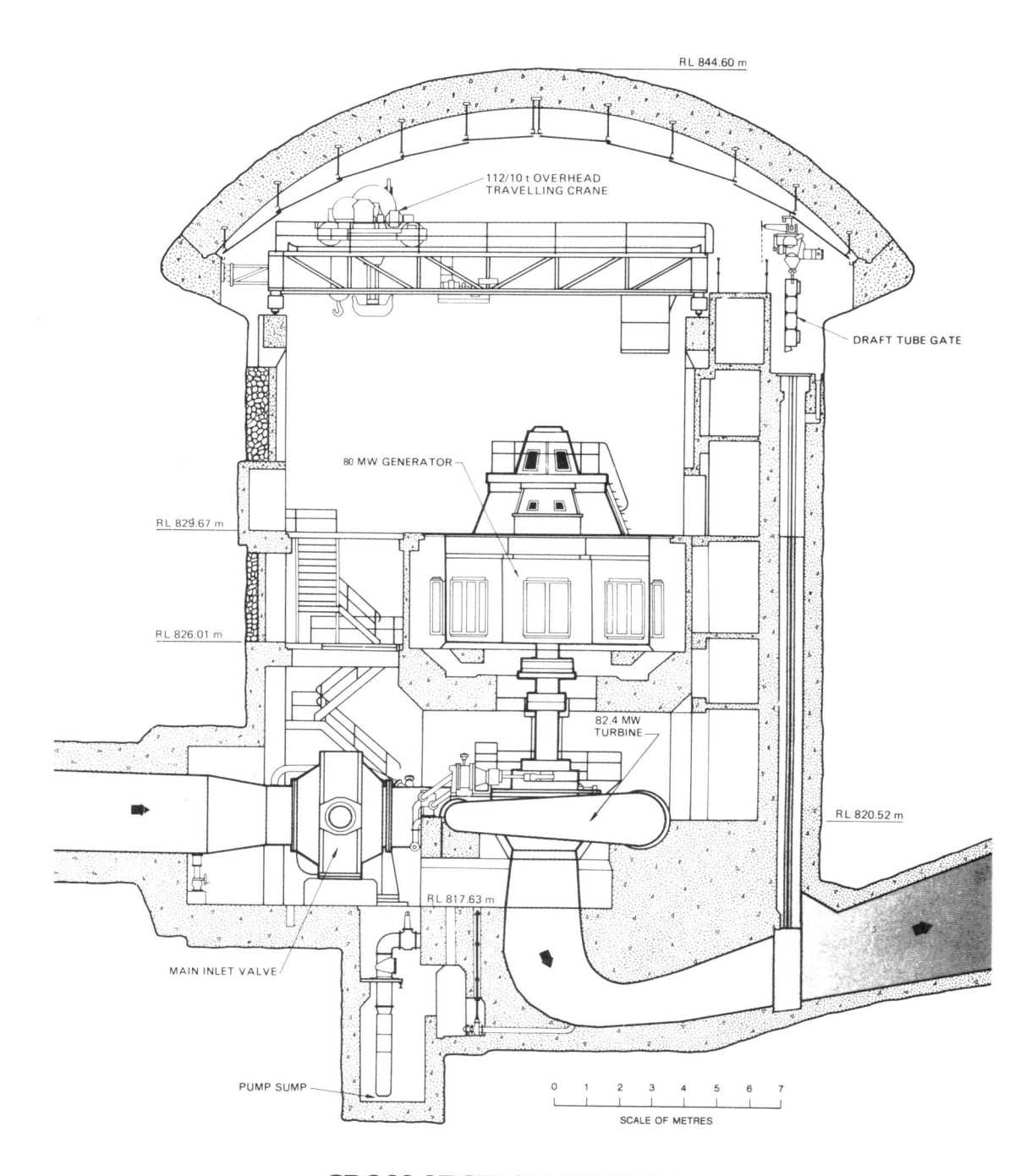

**CROSS-SECTION OF TUMUT 1
UNDERGROUND POWER STATION**

TECHNICAL INFORMATION

The machine hall of Tumut-1 underground power station at the completion of construction, 1959.

ISOMETRIC VIEW OF TUMUT 1 POWER STATION

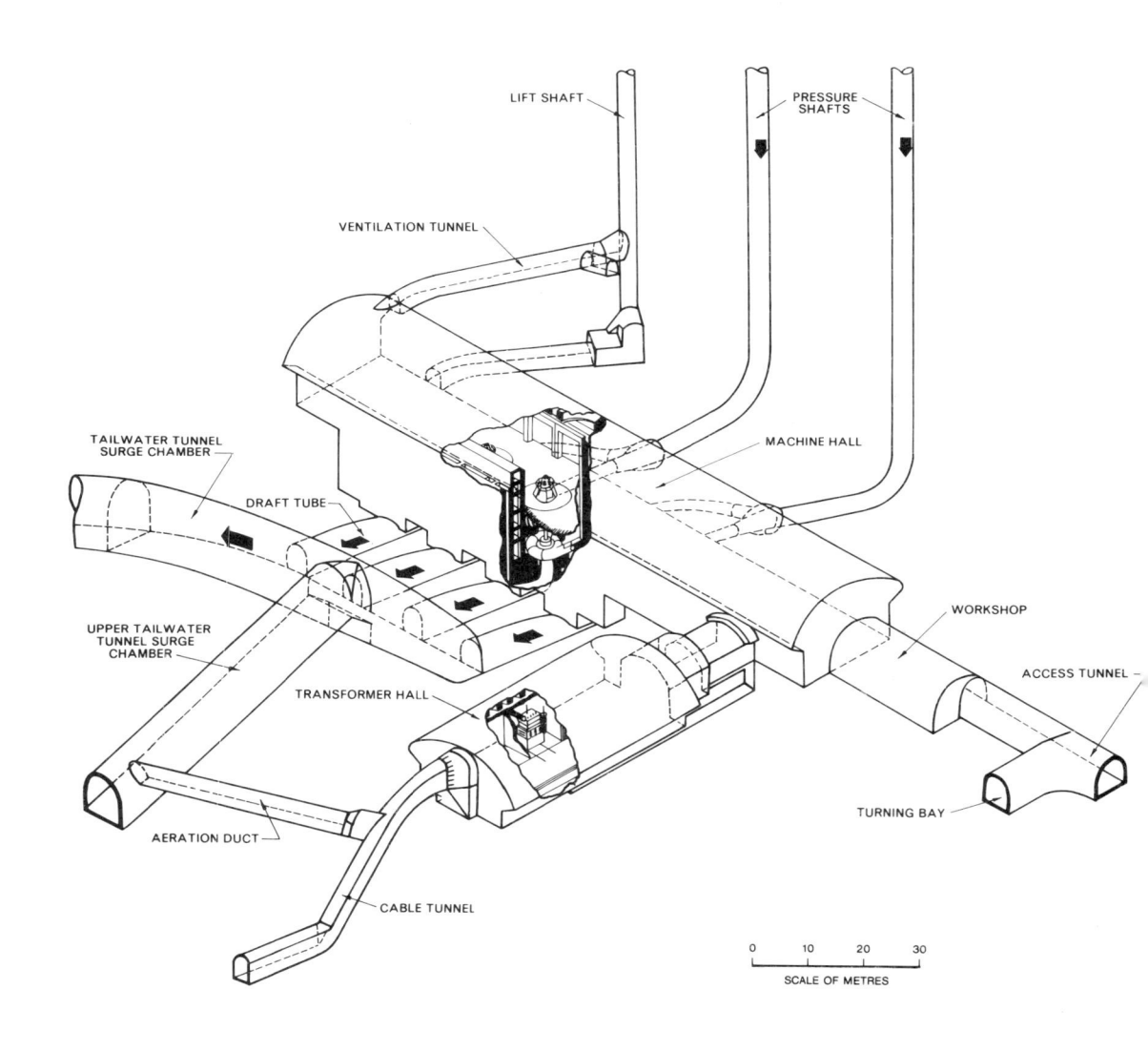

TECHNICAL INFORMATION

PLAN OF EUCUMBENE–TUMUT TUNNEL

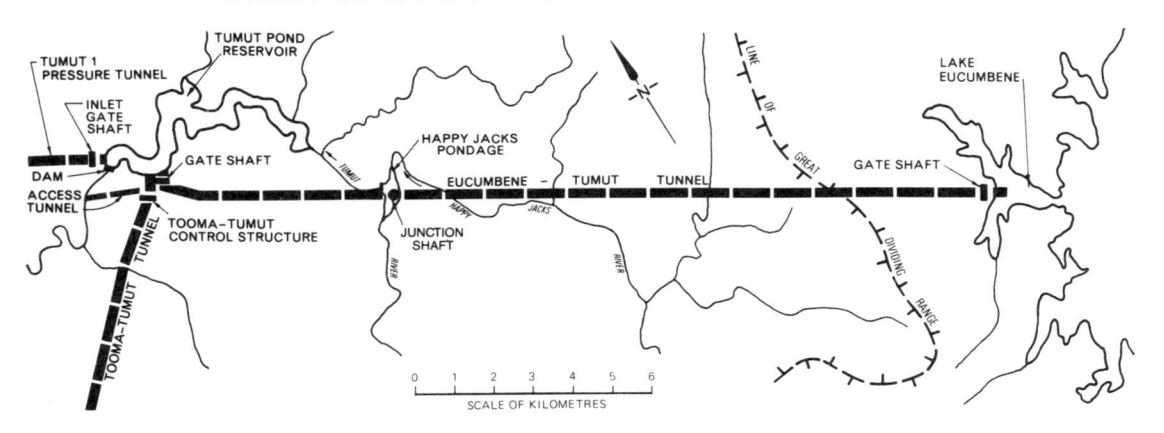

PROFILE OF EUCUMBENE–TUMUT TUNNEL

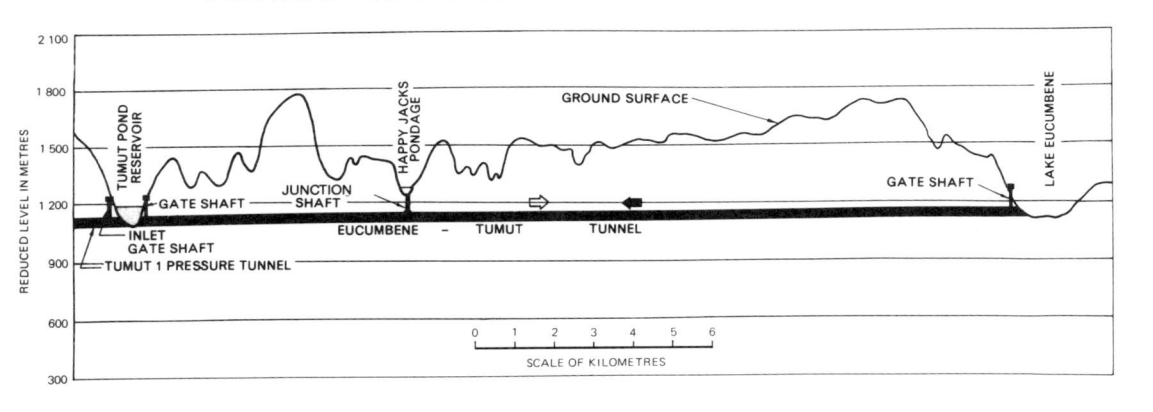

TUNNEL CROSS-SECTIONS

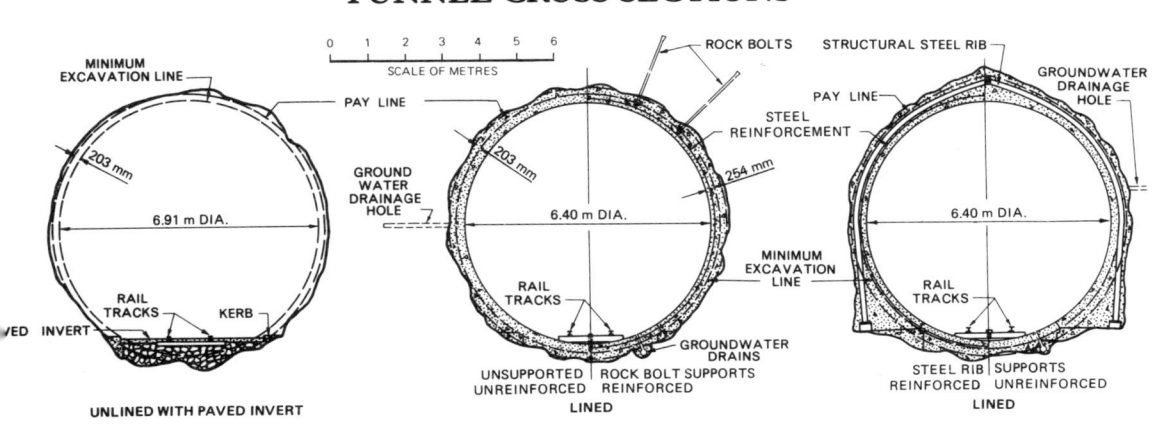

SNOWY-TUMUT DEVELOPMENT

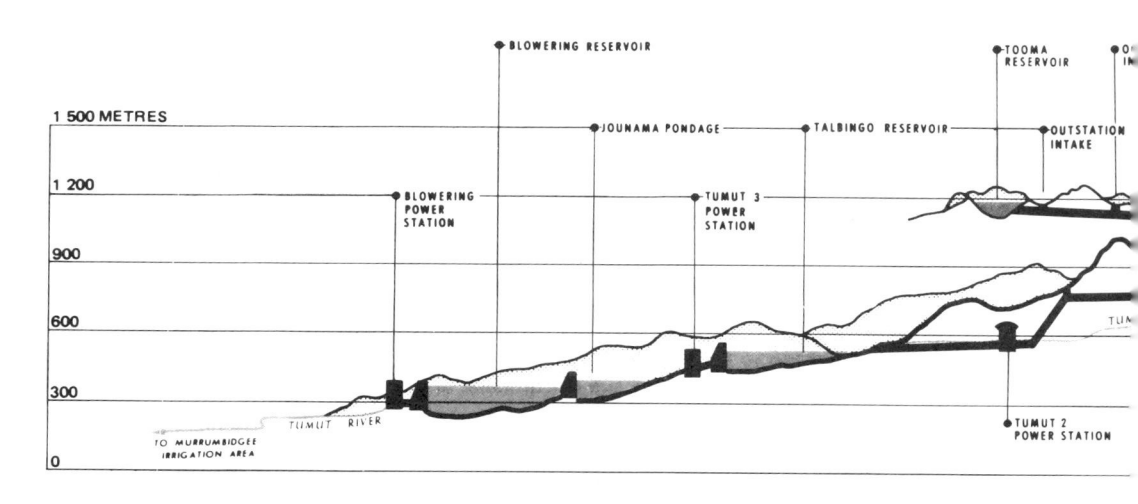

SNOWY-MURRAY DEVELOPMENT

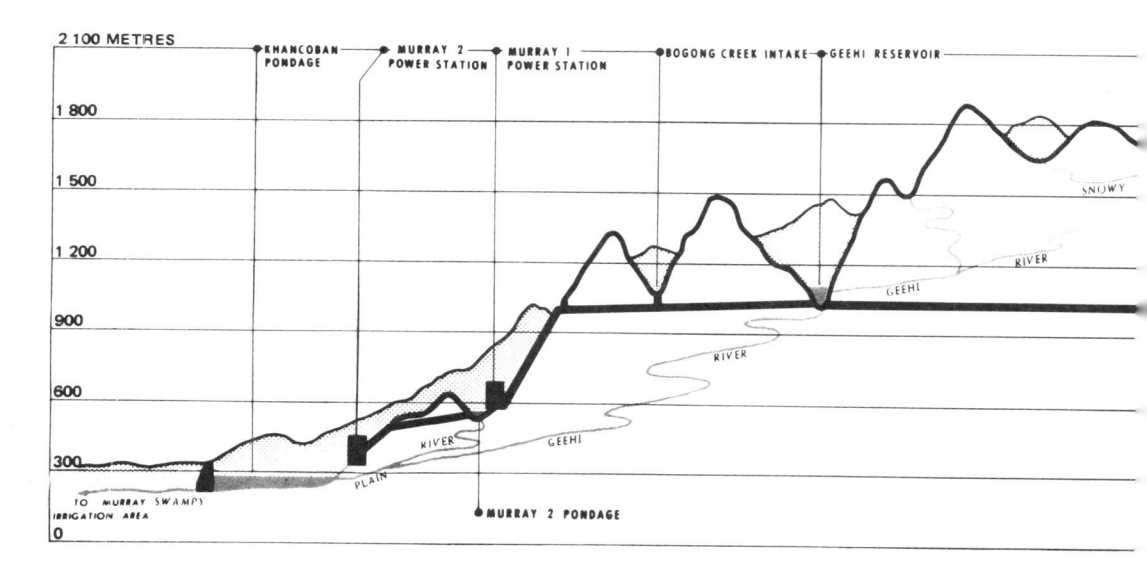

TECHNICAL INFORMATION

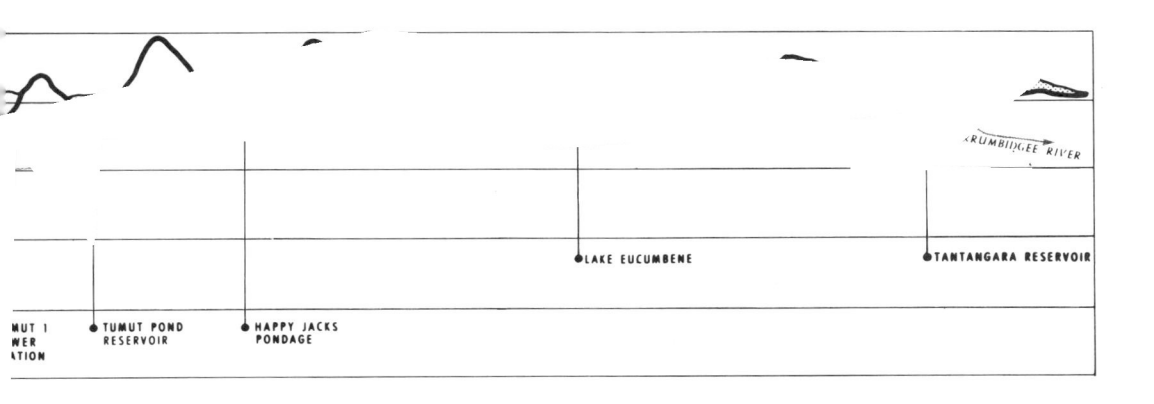

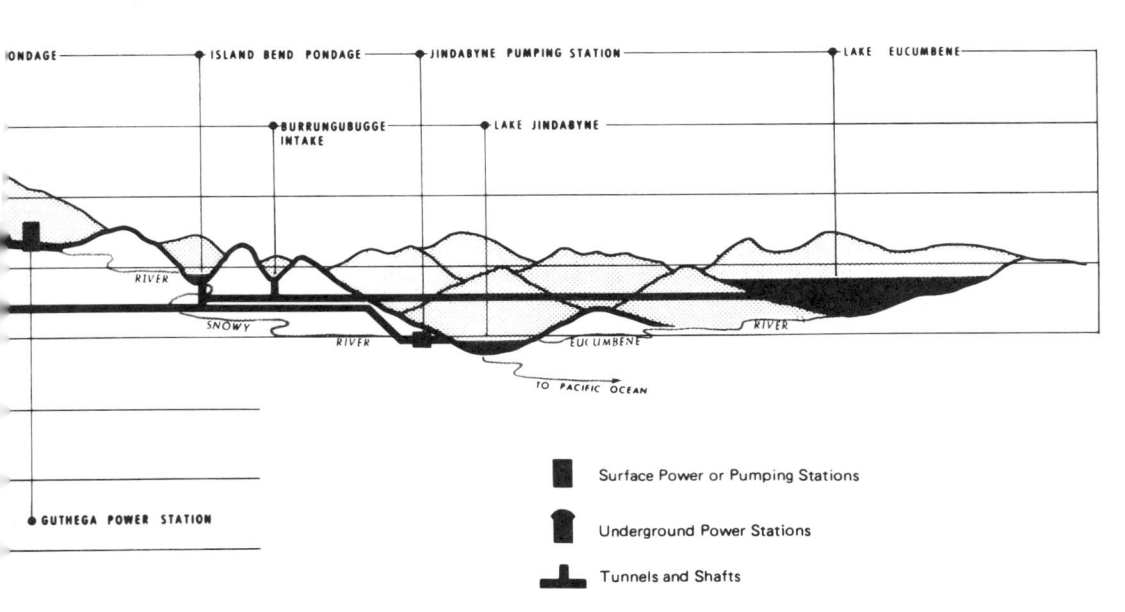

- TUMUT POND RESERVOIR
- HAPPY JACKS PONDAGE
- LAKE EUCUMBENE
- TANTANGARA RESERVOIR
- MURRUMBIDGEE RIVER

- ISLAND BEND PONDAGE
- JINDABYNE PUMPING STATION
- LAKE EUCUMBENE
- BURRUNGUBUGGE INTAKE
- LAKE JINDABYNE
- SNOWY RIVER
- EUCUMBENE RIVER
- TO PACIFIC OCEAN
- GUTHEGA POWER STATION

■ Surface Power or Pumping Stations
▆ Underground Power Stations
⊥ Tunnels and Shafts

F.S.L. Full Supply Level   M.O.L. Minimum Operating Level
R.L. Reduced Level (Elevation in metres above sea level)

INDEX

Names of people mentioned in the book (figures in *italic* type indicate photographs):

Ackerman, P., 287
Adams, Garry, 287
Adams, Jack, 77–78
Adams, P.F., 35
Amelung, Dieter, *46,* 64–66, *67,* 119, *120,* 121, *188, 195,* 195–98, *196, 197,* 209–10, 273
　see also colour section
Amelung, Heather, 209–10
Andrews, Ken, 84, 146, 263, 276
　lining tunnels, 132
　seatbelts introduced, 92, 94
Archer, Bob, 84
Argyroupoulos, C., 287
Arnold, Kurt, 98
Ashnikov, Igor, 208

Baksa, Dr Jon, 181, 185–86, 201
Balde, 223
Balzer, W., 286
Baragry, Edmund Joseph, 70
Barrow, Jeff, 288
Beard, Nicholas J., 287
Beccia, L., 287
Bein, Dr, 151–52
Bennett, George, 177
Berents, Derek (jr), *159*
Berents, Derek (sr), 156, 157, 161
Berents, Dr Ina, 155, 156, *159,* 161–62, 182–83
Berents, Harold, *159*
Berg, J., 286
Bergerhausen, Hein, *51,* 60, 102, 103–4, 105, *277*
　after leaving the Authority, 276–77
　early days in Cooma, 98–100
　war experience, 47–58
Bergerhausen, Sybille, 47, 52–53, 55, 57, 60, 276, *277*
　early days in Australia, 102–3
Big Romeo, 222, 223
Big Willie, 185
Bird, Gordon, 223–24

Birnbaum, Karl Heinz, *see* colour section
Biscontin, Tomaso, 287
Bladen, Ken, 226
Blaxland, Gregory, 34
Blhem, Del, 207
Blissett, Bob, 217
Boasso, Father, 158
Bolte, Sir Henry, *174, 175,* 176
Bolton, Dudley, 253, 254
Bolton (née Crowe), Dolly, 253, 254
Bott, Serafino, 288
Bowen, Joe, 179, 181
Brander, N.L., 287
Breier, Christian, 288
Brenycz, D., 286
Brescacin, Elio, 288
Briska, J., 288
Brydl, Dusan, 151–52
Buckman, H.J., 286
Bullock, Neville J., 288
Burns, David Allen, 288

Cacic, N., 287
Cahill, Joe, 80, *174, 175,* 176
Calwell, Arthur, 110
Cambell, Jon, 207, 208, 273
Carter, Noel, 136–37
Casey, Richard Gardiner, 85
Casson, Aub, *221*
Cavanagh, Sergeant, 153, 155
Cerubino, L., 286
Chapman, Fred ('Careless Hands'), 213–14, 217, 218, *218*
Chifley, Ben, 8, *32,* 35, 42, 80, *81,* 85, 138, 277
　accepts European refugees, 31, 33
　broadcasts support for the Scheme, 37–38, 281
　political support for Lemmon, 198
Cleary, Jon, 254
Clempson, Brian, 223
Clews, Major Hugh Powell Gough, *16,* 17, 20, 29, 76, 82, 94, *94,* 95
　early work on the Scheme, 24–28
　in retirement, 272–73
　see also colour section

INDEX

Compton, E., 288
Coranov, Jere, 184, 288
Costello, 287
Covic, Peter, 288
Crowe, Bill & Freda, 253, 254
Crowe, Tom, 253
Cupido, Franchesir, 287
Cutting, Bill, 215

Daffy, John, 84
Dahlberg, Graham E., 287
Dall' Antonia, Silvo, 286
Dan, Howard, 272
Darsie, Ivo, 186, 287
Davey, Ron ('The Admiral'), 217, *218*, 218–19
Dawson, A., 287
Degani, V., 286
Delacombe, Lady, *212*
Delacombe, Sir Rohan, *212*
Del Dorso, Carlo, *159*
Del Santo, Santo, *159*
Derrig, William, 288
De Vries, H., 287
Diatchenko, Nickolas, 288
Di Salvio, M., 287
Dramisino, S., 287
Dreise, Bob, 90
Duane, Patrick J., 287
Dumitrascu, Con, 163–64
Dunbar, Whitey, 181
Durrant, Bruce 'Bags', 217, *218*
Dykes, Norman, 70
Dykes, Stan, 70

Egan, Michael, 288
Eggeling, Bert, 17, 82
Elizabeth II, Queen of England, 183
Etheridge, Harold, 217, *218*
Evans, Jack, 286

Fabbro, Alfredo, 116–17, 155–56, 157–58, *159*, 276
Fabbro (née Miane), Guiditta, 116–17, 155–56, 157–58, *159*, 232, 276
Fabbro, Robert, 158
Fairbairn, David, 85, 270
Farrant, F.J., 286
Farrell, R.M., 286
Fazey, Thomas, 287
Fergusson, R., 287
Fizzoll, B., 287
Flood, Dr Josephine, 22
Foxall, Ian, 177
Franklin, Dr M.C., 36

Gaffy, Alan Thomas, 288

Gare, Neville, 263
Gasparetto, Dahilo, 287
Georg, Fred, 177
Gibbs, Frank, 113
Giesel, Whitney, 225
Gino, 184–85
Gleeson, 80
Graham, Bill, 213–14, 217, 218
Grazzi, Eno, *188*
Grigori, Peter, 288
Grodum, Olaf, 288
Gusela, M.A., 287

Hager, Joseph, 287
Hain, Herb, 77, 78
Hallop, Clifford, 287
Handler brothers, 78
Hann, Arthur 'Ossie', *218*
Harodes, Leonid, 186, 287
Hartwig, Walter, 58–60, 104, 135–36, 209, 270, 275
Harvey, Alf, 257, 258–59
Head, Bill, 217
Henriksen, N., 286
Henriksen, P.B., 286
Herzberg, Valker, 287
Hetherington, Alan 'Happy', 161
Heyerma, Hendrick, 288
Hinton, Joe, 84
Hirkic, Sharban, 287
Hiscox, Brian, J., 288
Hitler, Adolf, 47, 50, 53, 108
Holmes, Bill, 104, 107, 210–11, 217, *218*, 221, 226, 275
Holt, Harold, 135, 194
Hudson, Eileen, 44
Hudson, Sir William, 8, 40, *43*, 45, 83, 99, 114, *129*, 138, 142, 162, *180*, 198, 237–38, 263, *271*, 273, 278
  appointment as commissioner, 42, 44
  hard taskmaster, 41, 132, 140
  and Judge Taylor, 144
  problems and power struggles, 127–30, 134, 135, 136, 137
  retirement, 269–72
  seatbelts introduced, 92, 94
  segregates nationalities, 104, 199
  and Tony Sponar, 245
  unemployment surge, Cooma 1960, 247, 251
Humphrey, Arthur, 84
Hunter, Ron, *212*, 225

Ilic, M., 286
Ilic, Vladimir, 287

Jackson, Sir Robert, 31
Johansen, Ludwig, 287

Johnson, Alderman, 247
Johnston, Frank, 177
Jonjoric, Eddie, 286

Kaiser, Edgar, 207
Kajpust, Stan, 27, *73*
Kalvis, Albert, 27
Keefe, Bill 'Chubby', 217
Kerr, Deborah, 254
Kirsch, Henry, 177
Kobal, Ivan, 60, 62-64, *148*, 149-53, *151*, 275
Koch, G., 287
Kohler, H., 288
Kokles, Wayne A., 288
Kolari, S., 288
Kosciuszko, Tadeusz, 11, *14*, 15, 34
Kouchouvelis, Peter, 288
Kovacevic, L., 288
Kraljevic, Tomislav, 225
Kulis, Mirko, 225
Kurent, E., 288

Lafuente, A. (Pedro), 184-86, 288
Lang, Thomas Arthur, 84-85, 127, *129*, 129-30
Lascukowski, H.W., 287
Lawson, Jack, 111-14, *112*, 136-37, 276
Lawson, Olive, 114-15
Lawson, William, 34
Leech, Bob, 110, 111, 261, 262, 276
Leech, Helen, 261
Leech, Professor Tom, 110, 111, 114
Lemmon, Nelson, 8, 35-36, *36*, 42, 80, *81*, 277
  and environmental protection, 82, 280
  retirement from politics, 198
Lewis, Tom, 84, 263
Lomas, Kevin 'Lofty', 217
Longon, C., 287
Lucas, Gordon C., 288
Luisotto, M., 287
Luscombe, 258

McAlley, Eric, 286
Macarthur, Captain John, 11
Macarthur, James, 11, *12*, 14, 15, 34
McDonald, John, 287
McGufficke, Harry, 77
Mackay, Morris, 253
McKell, Sir William, 79, *81*
McMillan, Bob, 260
McQuade, Georgina, 205-6
Maier, Karl, 107-9
Mangelle, Joseph, 287
Maniatis, Peter, 287
Mansfield, Ross, 72
Martynow, Hildegard, 273
Martynow, Kon, 25-27, *73*, 177, 273
Mattner, Betty, 204-5

Mattner, Dick, 204-5
Mayorol, Aurelio, 184, 288
Menzies, Sir Robert, *39*, 80, 128, 138, 247
  opens Guthega Power Station 1955, 196-98
  opinion of the Scheme, 38, 45, 85
  retires as Prime Minister, 270
  see also colour section
Merrigan, Tony, 85, *129*
Mezlyak, Smail, 184, 288
Miniutti, Marcello, 287
Miraliotis, Nicholas, 288
Mitchum, Robert, 254, 255
Mori, Bogemir, 287
Mouat, Roger, 273
Mould, Boyd, 74, 75, 76, *246*, 259-63, 275
Mould, Ian, 124, 125, 126, 127
Mould, Merle, 202, 204
Mould, Reg, 75, 76
Mueller, Bill, 177
Munro, John 'Darby', 41-42, 84
Murphy, Dan, 87
Murray, Rod, 217

Nagy, J.B., 288
Nasielski, Adam, 201-2
Nasielski, Ksenia, 201-2, 276
Neely, Les, 206
Neithe, Len, 239-42, *241*, 276
Nekvapil, Sasha, 243, 245
Nemeth, L., 287
Newby, John, 150
Newell, Allan, 88
Nichols, Eric, 245

Oakley, Daniel, 288
O'Boyle, Ulick, 184-86, 273
Old Percy, 153, 155
Oliver, Charlie, 140, 142, 144, 163
Olsen, Olaf, 84

Pagge, Alfonso, 287
Palumbo, Franco, 149-50
Patchet, Harry, 153, 155
Paterick, Bill (jr), 122
Paterick, Bill (sr), 76
Paterick, Max, 76-77, 94-95, *123*, 275
  and danger in tunnel, 121-23
  rescues snowbound men, 237-39
  and shaft accident, 182, 183
Paterson, Andrew 'Banjo', 256
Patterson, Jimmy, 77
Patton, Don, 207-8
Pellegrino, D., 287
Pellegrino, Nardino, 72
Perez, Augustin, 288
Perkovic, D., 287
Phee, Neville, 206-7, 276

INDEX

Pighin, Angelina, 115–16, 155–57, 161, 276
Pighin, Mario, 115–16, 155–57, *159*, 161, 276
Pighin, Nadia, *159*
Pinazza, O.V., 286
Pinkerton, Iver, 84
Purcell, Colin, 169–70, 178, 179, 276
Puzey, C.P., 192

Reid, Alexander, 286
Reid, Don, 216
Ricardus, Peter, 177
Rieder, Walter, 288
Roden, Jack ('Kiwi Jack'), 161–62
Rodwell, Frank, 214–17, *215*, 225, 276
Rolling, Trevor, 41
Ronalds, Albert, 84
Rose, Cliff, 77
Rose, Professor A.J., 279
Rowntree, E.F., 84
Rudel, Hans Ulrich, 53
Rugolo, G., 287
Rullis, Peteris, 286
Rymal, Harry, 213–14
Rymere, Gert, 239–41

Satral, Louis, 288
Savegnano, Luigi, 287
Sbremac, Dusan, 73
Schafer, Bill, 206, 207
Scott, A., 23
Scott, W.W., 288
Seaman, N., 287
Sellan, Angelo, 286
Sergeant, Ian, 84
Sharpe, Ken, 153, *154*, 155
Shellshear, Wally, 136
Shultz, Brian, 210, 217
Signarini, Mario, *159*
Simpson, Bill, *128*
Sledzianowski, H., 287
Smolenski, M., 286
Sonter, Thomas, 226, *227*
Spencer, Charley, 75
Spencer, James, 75
Sponar, Tony, 243–45, *244*
Spooner, Senator William, 127–29, *128*, 130, 134, 140, *141*, 142, 270
Sramz, E., 288
Strzelecki, Paul, 11, *13*, 14–15, 273, 280
Svoboda, K., 288
Symonds, R.M., 286

Tarmo, Kalev, *230*, 231
Taylor, Mr Justice S.A. 'Stan', 140, 143–45
Texic, Mirko, 288
Thayne, D., 287

Thiess brothers, 146
Thiess, Kenneth A., 287
Thiess, Sir Leslie, 146
Thomas, 'Taffy', 210
Tsarevich, Milisov (Jacky), 73, *73*, 74
Tuite, T.B., 286
Turno, Aleksandryna, 15
Tyson, James, 34

Ulseth, R., 286
Ustinov, Peter, 254

Van Hoof, Peter L., 287
Vecchiato, Drenesto (Ernie), 187, 287
Vettor, Giuseppe, 287
Victoria, Queen of England, 15
Vyner, Robert, 23

Wagner, B., 286
Wahley, Carlin, 207
Waldon, Frederick, 186, 287
Waldren, John, 84
Wales, Bev, 117, *212*, 217, *221*, 222–26, 273, *274*
    and finding of *Southern Cloud*, 226–28
    new constable at Happy Jacks, 219–21
Wales, Rod, 223
Walsh, Darcy, 84
Walters, George, 287
Warburton, Dr Joe, 284
Wassermann, Wally, 177
Waters, Norm, 88
Watts, John, 288
Webster, Les, 112
Weir, Neville, 222
Wentworth, William Charles, 34
Werner, Henry, 177
Whitlam, Gough, 198
Whitney, Lionel, 217, *218*
Wialletton, Alessandro, 164–65, 169, 170, 171–72, 275
Wigmore, Lionel, 42
Williams, Peter, 177
Wilson, Jock, 92, 153, 155, *212*, 225, 273
    arrival on the Scheme, 87, 88, 90
    snowbound, 199–200
    see also colour section
Winnicombe, J., 287
Winston, Professor Denis, 252
Woodwell, Reverend Frank, 247, 251
Wright, Mr Justice S.C.G., 140
Wysocky, Edward, 25–27

Yannacopoulis, T., 286
Yong, N., 287
Yule, Gordon, 288

GLOSSARY OF ABORIGINAL NAMES

ADAMINABY	Resting/camping place
ADELONG	Plain with a river
BEGA	Big camping ground
BERMAGUI	Resembling canoe with paddles
BIBBENLUKE	Big lookout
BOGONG	A large moth
BOLARO	Man with boomerang
BOMBALA	Meeting of the waters
BUNYAN	Place of pigeon
CANBERRA	*Canberri* or *Nganbirra*; Meeting place
CHAKOLA	Lyrebird
COOLAMATONG	Water near a hill
COOMA	Big lake/Open country
GUNDAGAI	Going upstream/Sinews
JINDABYNE	Valley
JINDALEE	Bare hill
KIANDRA	Sharp stones used for knives
KUNAMA	Snow
MITTAGONG	Little mountain/A companion
MONARO	*Maneroo*; Treeless plain
MURRUMBIDGEE	Track goes down here/A very good place/Big water
NIJONG	Water
NIMMITABEL	Source of many streams
NUMERALLA	Valley of plenty
ORANA	Welcome
TALBINGO	Big Belly
TUMBARUMBA	Place of big trees/Hollow sounding ground
TUMUT	By the river
YARRANGOBILLY	Flowing stream
YASS	Waters